"AI" MEANS FRAUD:

THE FRAUDULENT REDEFINITION THAT'S DRIVING A GLOBAL CON

Jared A. Snyder

Synesi
Publishing
An imprint of Zoe's Publishing, LC

Synesi Publishing
An imprint of Zoe's Publishing, LC
Wyoming, USA

Disclaimer

The information presented in this book reflects the informed analysis, technical conclusions, and personal opinions of the author, based on over three decades of relevant experience. While every effort has been made to ensure accuracy at the time of writing, the author and publisher disclaim any liability for errors, omissions, or outcomes resulting from the application of this material. Any references to organizations, platforms, technologies, or individuals are made solely for critique, commentary, or illustrative purposes. Nothing in this publication should be interpreted as an endorsement, nor as legal, financial, or professional advice. Readers are encouraged to think critically and draw their own conclusions.

This book is dedicated to the memory of my grandfather, Carl Wentz—a WWII veteran awarded the Purple Heart, a farmer who worked the land with true grit, a devoted husband, and a loving father. When I was just six years old, he shared a piece of wisdom that's stayed with me ever since: "In your lifetime, you will come to learn that everyone says they want the truth, but nobody seems to know what to do with it." Those words, spoken with the weight of a life shaped by integrity, honor, and courage have, unfortunately, been proven true time and time again as I've watched the world wrestle with deception—including the very "AI" con job I expose on the pages of this book. Yet, with this book, I hold onto a flicker of hope that, if even just once, I can prove my grandfather wrong. My prayer is that the truth I've shared herein, rooted in reason and faith, will not only be discovered but also acted upon, helping others to see through the lies and stand firm in the truth regarding so-called "AI" technology. To the memory of my late, beloved grandfather, Carl Wentz—may his legacy of courage and truth live on in this work.

Contents

Introduction

Section 1 — Why This Book Exists: Exposing the Deception Behind So-called "AI"

I—1—1

A good non-fiction book introduction should begin with the author clearly describing the reason the book exists. In the case of this book, it exists to expose the fraud and deception behind so-called "Artificial Intelligence," or "AI," as it's more commonly referred to. It's as simple as that. I'm not going to try to capture your attention with profound quotations or purely emotional sentiments—nothing like that. And unlike the technology industry and the marketing teams that are selling you a bill of fraudulent goods, I'm going to strip out all the fake hype and rhetoric and simply tell it to you straight—in as simple and concise terms as I can. Now, onto my thesis statement (because every good non-fiction book needs one): Over the course of this book, I will reveal how the term "AI" has been fraudulently redefined and exploited to propel a supposed modern technological revolution that is, at its core, nothing more than an international con job. I will also provide insight into what—in my best estimation—is driving this particular cycle of fraud from the technology industry and the far-reaching consequences it is causing.

With that out of the way, you'll want to know a little bit about who I am and what my real motivation for writing this book is. I am a technology professional with roughly three decades of experience. I have worked in or have a respectable knowledge of—I'd dare say—nearly every corner of the technology world, especially Information Technology (IT). Moreover, I have extensive experience working firsthand with the technologies that make up today's so-called "AI." I've seen the industry grow and evolve over a long period of time, and I have witnessed the hypes and blatant hoaxes it has produced over the years— including the infamous Y2K Bug, which was supposed to bring about the collapse of modern civilization as we know it. Prisons were supposed to spring open, and aircraft were supposed to fall from the sky. Now, of course, I didn't buy into the manufactured panic for even one minute. I can even recall attempting to quell my dad's anxiety by reconfiguring the date and time on our Tandy 1000 TL/2—the first computer my family ever owned— as well as on our newer IBM Aptiva, setting them both to January 1, 2000, 12:01 AM, rebooting the machines, and showing him that they still both worked just fine. He ended up buying a generator and stocking up on food and water anyhow. *C'est la vie.* Now to be clear, I'm not saying there wasn't anything to the Y2K Bug claim, but the actual bug was largely tied to old COBOL systems in banking and government that used two-digit years, like "99" for 1999, which could've misread "00" as 1900 instead of 2000, throwing off things like interest calculations or record dates. What I am saying, is that the severity and potential reach of the bug were incredibly over-exaggerated for the sake of profit. As such, with all the fear muddying the water, the billions spent on "Y2K readiness" could never be fully attributed to only necessary repairs or upgrades, leaving one to wonder how much of it was driven by panic and opportunism, with tech companies cashing in on all of the hype.

I—1—3

Fast forward 25 years, and I am watching the world being conned by the technology industry once again—this time with fraudulent narratives and marketing campaigns boasting of "Artificial Intelligence" technology. As much as the over-exaggeration of the Y2K Bug annoyed me then, the present "AI" scheme frustrates me all the more, largely because of the numerous hats I now wear. In addition to being a technology veteran, I am also a Christian leader and mentor—known to some as Pastor J. So, apart from just knowing the truth about the current "AI" narrative—as a technology expert—I now also maintain that I have a faith-oriented obligation to ensure that it is revealed for the fraudulent nonsense that it is. Between my extensive knowledge, my commitment to truth, and my ability to conduct thorough research to fill in any gaps, I am confident that, by the end of this book, readers will clearly understand why the "AI" narrative is a calculated and deliberate fraud designed to manipulate public perception and drive profit. More importantly, they will gain the discernment needed to navigate this technological era with wisdom, truth, and a firm grasp on reality.

Section 2 — What People Are Led to Believe vs. Reality: Addressing the Myth of "Intelligence"

I—2—1

I would be willing to assert that, for many of you, when you hear the term "Artificial Intelligence," you likely envision technology—of some form—capable of reasoning and making decisions in a way that mirrors how human beings do the same. By and large, you would not be wrong to expect this. In fact, such an expectation largely aligns with the original intent behind the concept of "Artificial Intelligence." The problem is not with your understanding of what "AI" is supposed to entail, but rather with the narratives and marketing being pushed about the technologies currently labeled as such. For years, promotional narratives surrounding so-called "AI" have conditioned the public to believe that we already have—or are on the verge of achieving—machines that can truly think. This portrayal has become so ingrained in public consciousness that many people—perhaps even you—now accept it as unquestioned fact. However, the real nature of these technologies is quite different.

I—2—2

The reality of modern so-called "AI" technologies is that they are far from intelligent—artificially or otherwise. They cannot reason, they cannot feel (not that they're supposed to), and they do not possess consciousness, self-awareness, or any other property that constitutes genuine intelligence. Moreover, they cannot understand language nor underlying concepts—basic characteristics the definition of "Artificial Intelligence" requires. What's more, far from being autonomous—meaning capable of making informed, independent decisions—these systems rely entirely on complex algorithms paired with large-scale data processing. They do not "think" or "understand"; they generate responses that appear correct or appropriate only because they process vast amounts of data through statistical models. Put simply, they are not reasoning—they are gambling, predicting what the most statistically probable response should be based on their programming and training datasets. At their core, they are doing exactly what computers have always done: compute—only now at a much larger scale.

I—2—3

The distinction between what people *believe* to be intelligence and what these technologies are actually doing lies at the heart of the "AI" hoax. The main perpetrators of this fraud—along with much of the technology industry at large—have deliberately blurred the lines between computation and true intelligence to create an illusion of awe and wonder. Yet the reality falls far short of the spectacle they sell. When I think of the absurdity of it all, I cannot help but recall a scene from *The Wizard of Oz*, where Dorothy's little dog Toto pulls open the curtain to reveal not a mighty sorcerer, but a frantic old man yanking levers and turning knobs to maintain the illusion of a great and powerful wizard. Likewise, by invoking the word "intelligence," these companies have convinced the world that they are building machines that can think for themselves—but that is simply not the case.

I—2—4

At best, the technologies being labeled as "AI" today are nothing more than powerful—and resource-hungry—tools designed to perform specific tasks—often more efficiently than humans —while allowing for natural language input and output. These are not new capabilities; computers have been performing similar tasks for decades, albeit once requiring specialized skills to operate. So-called "AI" is neither a groundbreaking leap in technology nor a revolutionary new application of existing systems. It is simply a combination of older technologies, repackaged and rebranded to create the illusion of progress. More importantly, it is not —and will never be—intelligent in any human sense of the word. Understanding this fundamental distinction is essential for seeing through the myth of "AI" and recognizing the technology for what it truly is: sophisticated but inherently limited tools— tools that, despite the public's growing faith in their perceived infallibility, are still incredibly riddled with flaws.

Section 3 — The Role of Marketing in the "AI" Boom: How Companies Manipulate Perception

I—3—1

You may be wondering where the current "AI" boom came from—a question that everyone should be asking. I'll address this in greater detail later in the book, but briefly, the frenzy surrounding "Artificial Intelligence" has been fueled entirely by fraudulent—yet highly effective—marketing tactics. These campaigns are designed not only to capture attention and generate excitement, but also to spur massive investments and sell products and services. At the heart of this deception is a deliberate manipulation of public perception—a carefully orchestrated mix of buzzwords, lofty promises, and deceptive imagery, all aimed at convincing people to believe in something far more advanced and capable than these technologies actually are.

I—3—2

One of the most critical aspects of this manipulation is the deliberate misuse of the term "Artificial Intelligence." As I've mentioned, the word "intelligence" is inherently tied to an ability to understand language, grasp underlying concepts, and make uncoerced decisions. It evokes images of sentient machines capable of independent thought or autonomous behavior—like the robots and artificial beings we've seen in science fiction for decades. Yet, the reality is far removed from these portrayals. As I've already stated, these technologies do not reason; they merely predict the most likely response based on their programming and the vast datasets they've been trained on. In other words, they do exactly what computers were designed to do: compute—just on a much larger scale. The term "Artificial Intelligence" is used because it evokes a sense of wonder and capability—something that more accurate descriptions, like "statistical modeling" or "pattern recognition," couldn't achieve. However, this does not make the fraudulent use of the term any more acceptable.

I—3—3

Still, it is this fraudulent branding, combined with science-fiction imagery, that largely fuels the illusion of technological progress. The major players behind so-called "AI" technologies are fully aware that the public's imagination has long been captivated by the idea of intelligent technology and sentient machines. Nowhere is this more apparent than in Disney's Tomorrowland—a futuristic, tech-inspired section found in every Magic Kingdom across the globe. For decades, Disney has used Tomorrowland to evoke visions of a high-tech future, populated by autonomous machines and other dazzling technologies. Although these imagined technologies were once purely fictional, their persistent presence in the public consciousness continues to shape our expectations of real-world "AI." This cultivated sense of wonder —once driven by fantasy and now marketed as a reality—makes it easier for companies to manipulate the public into believing that today's "AI" technologies are far more advanced than they actually are, and even sentient. Just as Disney's futuristic landscapes were once pure fantasy, today's so-called "AI" technology is packaged in a similarly fantastical form—though it is not designed to challenge the imagination, as Disney's Tomorrowland did, but to make it believe in something far beyond the reality of what actually exists.

I—3—4

In reality, these technologies are far from intelligent. As I keep indicating, they don't reason, understand, or think independently; they operate through statistical analysis of data, not any genuine cognitive process. However, by labeling these tools as "intelligent," marketing teams have managed to convince the public—and, crucially, investors—that these systems can make decisions or perform tasks autonomously. In truth, they're simply executing pre-programmed instructions on a much larger scale than ever before. To make matters worse, there's the constant overpromising of capabilities. Companies often claim their "AI" systems can solve complex problems, automate entire industries, or even replace human workers. These bold promises capture headlines, generate interest, and attract investments. Yet when these systems fail to live up to their claims, the goalposts are simply moved. A system that can't meet expectations today is rebranded and repackaged, with new promises for tomorrow. After all, the next big thing in the "AI" revolution is always "just around the corner"—if only we invest a little more and wait a little longer.

Sadly, other key players—such as the media and politicians—also play a crucial role in perpetuating this fraud. Instead of critically examining the claims made by these companies, they often act as amplifiers, parroting corporate press releases and sensational headlines. Whether it's the fear of "AI taking over jobs" or the excitement of "AI revolutionizing healthcare," these narratives only help fuel the myth of "AI" as a transformative force. Rarely do we hear the truth —that what's being called "AI" is actually just advanced tools that can't think, feel, or reason the way humans do, nor even to the extent that the definition of "artificial intelligence" requires. Without genuine scrutiny, the myth becomes accepted as truth. At its core, this marketing-driven hype isn't about advancing technology or benefiting society; it's about preserving an illusion that sells products, services, and—most importantly—attracts investment. The tech industry thrives on overpromises, exaggerated claims, and the manipulation of public perception. The more people buy into the illusion of "AI," the more money flows into its development. Recognizing that "AI" is far from intelligent—just sophisticated, but ultimately limited, tools—is essential for seeing through the myth and understanding such technology for what it really is:

a repackaging of existing technologies, not a groundbreaking innovation.

I—3—6

The truth here is unavoidable: what is being sold as "AI" is nothing more than a carefully crafted illusion. In the following chapters, we will break down this deception, expose the truth behind the marketing, and show how the tech industry has been manipulating the public for its own benefit. It's time to do away with the smoke and mirrors and understand these technologies for what they really are.

"It is impossible for a man to learn what he thinks he already knows."

Epictetus
Discourses, Book II, Chapter 17, 108–125 AD

"The Analytical Engine has no pretensions whatever to originate anything. It can do whatever we know how to order it to perform."

Ada Lovelace
Notes on Menabrea's Paper, 1843

Chapter 1:
The Origins of the "AI" Concept (Historical Context & Early Overpromising)

Section 1 — Alan Turing and the Imitation Game (1950): The Theoretical Groundwork for Machine-based Intelligence

1—1—1

To truly grasp the origins of what is now called "Artificial Intelligence," we must go back to 1950 and examine the work of Alan Turing. As one of the pioneering figures in computer science, Turing is often credited with introducing the idea that computers might one day replicate aspects of human cognition. In his seminal paper, Computing Machinery and Intelligence, he posed a question that would shape decades of debate: "Can machines think?" However, Turing quickly recognized the difficulties of answering this question definitively, as defining "thinking" in a meaningful way proved to be elusive and philosophically challenging. Instead, he proposed an alternative approach, suggesting a test that would focus not on whether machines could think, but whether they could imitate human behavior convincingly enough to fool an observer. This shift led to the development of his *Imitation Game*—now widely known as the *Turing Test*—a concept that continues to influence discussions of so-called machine intelligence today.

This experiment, in essence, was Turing's attempt to sidestep the question of whether machines could truly think by reframing the issue around imitation rather than cognition. The test was straightforward: a human judge would engage in a text-based conversation with two unseen participants—one human, the other a machine. If the judge failed to reliably distinguish between the two, the machine could be said to have successfully imitated human responses, creating the illusion of intelligence. However, it's crucial to note the key distinction—illusion of intelligence, not actual intelligence. Furthermore, Turing's reasoning contained a critical flaw: he equated intelligence with the ability to deceive, rather than with true understanding or comprehension. The ability to fool a person into thinking they were conversing with another human doesn't prove that the machine possesses genuine intelligence. It simply means the machine excels at pattern recognition and response simulation. This distinction is paramount. Genuine intelligence requires reasoning, self-awareness, and comprehension— none of which are present in machines that simply regurgitate preprogrammed or statistically generated responses.

By Turing's logic, one could conclude that a ventriloquist's dummy, if operated skillfully enough to deceive someone into thinking it was alive, should be considered intelligent. Or a parrot, capable of mimicking human speech, should be considered just as intelligent as the person it mimics. The absurdity of such a conclusion is obvious—yet this is precisely the flawed reasoning embedded in the Turing Test. It did not measure intelligence; it measured a machine's ability to exploit human perception. Humans are naturally inclined to see patterns, assign meaning, and anthropomorphize things that exhibit human-like traits. The fact that a person believes they are talking to another human does not mean intelligence is present; it simply means the machine has effectively managed to manipulate expectations. This, of course, is a logical bait-and-switch—replacing actual intelligence with mere imitation—and it's precisely what the modern tech industry has built its "Artificial Intelligence" narrative upon. The majority of companies marketing so-called "AI" today are not just claiming that their systems mimic human behavior; they actively present their systems as intelligent, reasoning entities capable of understanding and decision-making. These companies perpetuate the myth of "AI" by leveraging the illusion of intelligence to manipulate perceptions, banking on the fact that the average person will assume that anything that appears intelligent must, in fact, be intelligent.

1—1—4

Turing, in his time, was making both a philosophical and technical argument. While he likely did not intend for this conflation to become a tool for some widespread con, the tech industry has taken his flawed premise and weaponized it. They now use it to market systems that do nothing more than predict statistically probable responses, presenting them as something more—thus perpetuating a myth that has ballooned into a full-scale worldwide fraud.

Section 2 — John McCarthy and the Dartmouth Conference (1956): The Birth of the Term "Artificial Intelligence"

1—2—1

Now that we have Turing's early experiment on machine intelligence as a historical reference, we can jump ahead to 1956—a pivotal year in the history of "Artificial Intelligence". At Dartmouth College in Hanover, New Hampshire, a group of researchers led by John McCarthy, an American computer scientist and cognitive scientist, organized the Dartmouth Summer Research Project on "Artificial Intelligence". This event is widely regarded as the moment when "Artificial Intelligence" emerged as a formal scientific discipline. The goal was to explore whether machines could be made to simulate aspects of human intelligence. McCarthy, along with other distinguished participants such as Claude Shannon, Marvin Minsky, and Nathaniel Rochester, hypothesized that "every aspect of learning or any other feature of intelligence can in principle be so precisely described that a machine can be made to simulate it." This was the first clear articulation of "AI" as a scientific endeavor—one with the potential to be studied, advanced, and ultimately realized in machines.

<h2 style="text-align:center">1—2—2</h2>

The Dartmouth Conference is also credited with the formal introduction of the term "Artificial Intelligence." John McCarthy is often recognized for coining the phrase to describe the initiative aimed at creating machines capable of simulating— not mimicking—human intelligence. This event gathered experts from diverse fields—mathematics, neuroscience, psychology, and engineering—each bringing their unique perspective to the monumental challenge of constructing "thinking machines." The ambition was nothing less than to develop machines that could perform tasks traditionally associated with human intelligence— problem-solving, reasoning, learning—at a level not merely imitating but matching human capabilities. The Dartmouth Conference thus marked the formal birth of "Artificial Intelligence" as a research field, setting the stage for future innovations that would captivate the world's imagination.

1—2—3

Despite the excitement that surrounded the Dartmouth Conference, the early years of "AI" were filled with significant challenges. Although the term "Artificial Intelligence" was synonymous with the aspiration to replicate human intelligence, the technology of the time was woefully inadequate to fulfill such ambitious goals. While early "AI" researchers made notable progress with specific applications—such as solving mathematical problems or playing games like chess —the broader ambition of achieving human-level "AI" remained elusive. The initial optimism of the Dartmouth Conference gave way to a series of setbacks, as researchers encountered the deep complexities involved in simulating human cognition and problem-solving abilities. The ambitious vision of creating truly intelligent machines would have to wait, as the gap between hope and reality widened over the coming decades.

The Dartmouth Conference helped lay the groundwork for what would become a pervasive "AI win mentality" that still dominates the field today. This mentality is driven by the hope that machines might one day think, reason, and learn like humans. However, as many have come to realize, this hope is not only overly optimistic but also fundamentally unrealistic. The idea of creating "Artificial Intelligence" became a central motivator for decades of research, but it also set the stage for a recurring cycle of overpromising and underdelivering. The Dartmouth Conference marked the formal birth of "AI," but it also signaled the beginning of a long, frustrating journey, one marked by brief periods of perceived progress punctuated by deep stagnation and disillusionment. The gap between the vision of truly intelligent machines and the limitations of existing technology has been a consistent theme throughout the history of the field.

In much the same way that the Dartmouth Conference's predictions of imminent "AI" breakthroughs were overambitious, today's "AI" narratives continue to inflate expectations. We are frequently told that we are on the cusp of achieving true intelligence in machines, only to face the reality that current technologies are still grounded in statistical pattern recognition and modeling, not genuine reasoning or understanding. The same cycles of hype and disappointment that plagued the early days of "AI" continue to recur today, with each new technology—from so-called "deep learning" to large language models (LLMs)—being heralded as the next great leap, only to fall short in areas central to human experience: reasoning, adaptability, and true understanding. In nearly 70 years, despite some arguable technological advancements, "AI" has yet to bridge the gap between the promises made in 1956 and the limitations of what machines can actually do.

1—2—6

As "AI" research continued, McCarthy sharpened his vision in response to the growing recognition of the technology's deeper challenges. By 1969, he refined his original definition of "Artificial Intelligence" to emphasize a critical distinction: machines might be capable of exhibiting intelligent behavior, but this should not be confused with actual human intelligence. More importantly, McCarthy introduced the idea of "concept learning" into the definition—an idea that required machines to move beyond mere pattern recognition and statistical output. Rather than simply reacting to inputs or imitating intelligent behavior, machines would need to develop an internalized understanding of underlying concepts and relationships within the world they interact with. This refinement raised the bar significantly: a true "AI" technology would not merely predict or mimic but would have to demonstrate a form of conceptual grasp, akin to genuine reasoning and understanding. McCarthy's updated view exposed how shallow earlier interpretations of "AI" had been, while also signaling a shift toward a much more demanding—and as history has shown, far more elusive—goal for the field.

Section 3 — Early Optimism vs. Reality: Why Researchers Thought Human-like Intelligence was Imminent

1—3—1

In the wake of the Dartmouth Conference, researchers—still riding the wave of enthusiasm it generated—saw human-like intelligence as an engineering challenge rather than an insurmountable problem. Early "AI" programs, capable of solving logic puzzles, playing strategy games, and manipulating symbols, reinforced the belief that scaling up these methods would inevitably produce machines that could think, reason, and learn like humans. This assumption drove bold predictions, such as Herbert Simon's 1965 claim that machines would be capable of any human work within twenty years, and Marvin Minsky's 1967 assertion that human-level "AI" was less than a decade away. These proclamations epitomized the era's optimism—an optimism that would soon be exposed as disastrously premature.

1—3—2

Despite such confident predictions, the technological limitations of the time quickly exposed their flaws. Early "AI" systems operated within tightly controlled environments with well-defined rules—like board games and logic puzzles—but collapsed when faced with the ambiguity and complexity of the real world. Researchers had drastically underestimated the challenges of perception, flexible reasoning, and adapting to novel information. Machines could manipulate symbols, but they had no understanding of what those symbols represented. The assumption that scaling up symbolic logic would naturally lead to general intelligence proved to be fundamentally flawed. As progress stagnated, it became increasingly clear that human cognition was far more intricate than researchers had initially believed—and that symbolic "AI," built on rigid rule-based systems, could not bridge the gap between machines and anything even representative of genuine intelligence.

1—3—3

The inevitable result was a widening chasm between expectation and reality, with frustration firmly taking its toll. Government agencies and funding organizations, once eager to invest in "AI" research, grew impatient with the lack of meaningful results. By the mid-1970s, enthusiasm had decayed into disillusionment, as researchers found themselves trapped within the limitations of rigid, rule-bound systems. The inability of early "AI" to cope with real-world complexity triggered a significant decline in investment. What followed became known as the first "AI winter"—a period of stagnation during which interest, funding, and public confidence sharply declined. The earlier proclamations of imminent human-like intelligence had been revealed as little more than naive overconfidence, and the field was forced to confront a harsh reality: intelligence was far more complex than anyone had wanted to believe. The first "AI winter" lasted from the mid-1970s into the early 1980s, marking a time when many researchers were forced to fundamentally reconsider their approaches.

Section 4 — The "AI" Winters: Repeated Cycles of Overpromising, Underdelivering, and Funding Collapses

1—4—1

As the initial wave of enthusiasm from the Dartmouth Conference and the decades of inflated promises that followed began to lose momentum, reality finally caught up with "AI" research in the 1970s. Early systems—primarily symbolic and rule-driven—had shown narrow successes in domains like game-playing and basic problem-solving, but when faced with the complexity of real-world environments, they collapsed. Programs like SHRDLU, capable of manipulating blocks and carrying out simple tasks in a highly controlled setting, were hailed as major milestones in natural language processing. Yet SHRDLU, like its contemporaries, operated entirely within predefined boundaries and could not generalize beyond them. Despite the optimistic proclamations of researchers like Herbert Simon—who in 1965 declared that machines would rival human work capabilities within two decades—the field's fundamental limitations became increasingly apparent. As the 1970s progressed and "AI" repeatedly failed to deliver on its grandiose promises, funding began to evaporate. The U.S. Department of Defense, once a major backer, slashed its investments in 1973 after growing impatient with the lack of meaningful progress. The Lighthill Report, also published in 1973, delivered a harsh assessment of "AI's" stagnation and further fueled the retreat of financial support. Together, these events ushered in the first "AI Winter"—a period of disillusionment that lasted into the early 1980s.

1—4—2

The 1980s brought a new resurgence of interest, fueled by the rise of expert systems. These programs sought to encode human expertise into large rule-based knowledge bases to replicate professional decision-making in fields like healthcare, geology, and engineering. Systems like MYCIN, developed for medical diagnosis, demonstrated some initial promise, offering a glimpse into how "AI" could assist specialists. However, the limitations of expert systems soon became glaringly obvious. They were brittle, inflexible, and prohibitively expensive to develop and maintain. Updating knowledge bases to reflect new information was labor-intensive, and the systems themselves struggled to handle ambiguity or adapt to unfamiliar situations. As the promises surrounding expert systems once again failed to materialize, the second "AI Winter" took hold in the late 1980s and early 1990s. Funding dried up, research interest faded, and once-prominent "AI" companies like Symbolics, Inc.—founded to commercialize expert systems—collapsed under the weight of unmet expectations. Intelligence, as it turned out, could not be distilled into a static collection of rigid rules—a lesson that the industry, despite repeated failures, would continue to forget —or maybe intentionally ignore.

1—4—3

In the early 2000s, an altogether new wave of optimism emerged, fueled by advances in "machine learning" and the resurgence of "neural networks." In the wake of the Y2K Bug—an exaggerated crisis that had illustrated just how vulnerable and easily manipulated the tech industry could be—companies and researchers latched onto "machine learning" as the next big breakthrough. Computing power had grown significantly, and new algorithms allowed machines to achieve impressive results in narrow tasks like speech recognition, image classification, and natural language processing. Yet once again, reality lagged far behind the hype. These systems, though marketed as intelligent, were heavily dependent on vast amounts of labeled data, massive computational resources, and constant human intervention. Their perceived successes were confined to narrow domains; small deviations in input often produced unpredictable or absurd results. As the hype failed to match real-world performance, a partial funding pullback—a minor winter—set in. Although "machine learning" and "neural networks" kept "AI" alive in public imagination, the underlying cycle of overpromising, underdelivering, and disappointing investors persisted. This time, researchers and companies alike tried to temper expectations with a more cautious rhetoric, but the core technical limitations remained unresolved.

1—4—4

Despite the scars left by past "AI Winters," another surge—filled with more seemingly blind optimism—has emerged for "AI." While the technology industry should, by now, be wary of repeating its past mistakes, it instead charges ahead, once again making big promises—this time related to "deep learning"—a rebranding of techniques conceptually similar to earlier neural networks—as though the lessons of the past were forgotten, or perhaps intentionally disregarded. With traditional avenues of technological progress slowing, the industry has sought new frontiers to maintain momentum and justify continued investment. With traditional technological avenues slowing, particularly as Moore's Law reaches its physical limits, the industry has turned to "AI" to sustain its momentum. Moore's Law—an observation that the number of transistors on a chip would double roughly every two years—once propelled explosive growth in computing power.

However, as silicon-based processors approach physical limits, the exponential gains once expected from this law are slowing, reflecting the theoretical maximums of current technologies. In other words, it's like trying to fit more and more stuff into your luggage when it's already bursting at the seams— no matter how much you push, it's just not going to accept that extra jacket you want to pack, just in case the nights get chilly. This inevitable slowdown has forced the industry to explore alternative areas of innovation, and with quantum computing still far from being a viable technology, "AI" has been positioned—once again—as the next great technological revolution. However, if history is any indicator, the key players—including investors— had better bundle up warmly.

"God offers to every mind its choice between truth and repose. Take which you please, — you can never have both."

Ralph Waldo Emerson
Intellect (an essay), 1841

"Truth, like gold, is to be obtained not by its growth, but by washing away from it all that is not gold."

Leo Tolstoy
A Calendar of Wisdom, 1904

Chapter 2:
Moore's Law and the Stagnation of Real Technological Progress (Why the Tech Industry Needed a New Hype Cycle)

Section 1 — The Golden Age of Computing: How Exponential Transistor Scaling Drove Real Innovation

2−1−1

Of all the advancements in electronics technology over the years, the invention of the transistor in 1947 is arguably the most significant. It revolutionized the world of electronics, providing a new foundation for the development of computing. For those unfamiliar with the technology, at its core, a transistor is a tiny electronic switch that can control the flow of electricity. Think of it like a light switch: when you flip it on, electricity flows (turning the light on), and when you flip it off, the flow stops (turning the light off). In computers, transistors perform the same role by controlling the flow of electrical signals. In fact, a single modern computer processor might contain billions of transistors. For example, Intel's Core i9-11900K processor—a high-end chip used in desktop computers—contains about *19.2 billion transistors.* Each of these transistors switches on or off at incredibly fast speeds, controlling the flow of electricity through them, which in turn generates electrical signals that represent binary data as 1's (on) and 0's (off). When these transistors work together in coordinated patterns, they generate the binary code that serves as the fundamental language underlying all computer operations, enabling everything from simple calculations to complex tasks like running applications or browsing the Internet. Before the transistor, early computers relied on vacuum tubes, which served a similar function but were large, unreliable, and consumed significant amounts of energy. While vacuum tubes played a vital role in early electronic systems and continue to be useful in specific applications today, they were not well-suited for the demands of modern computing. Transistors, on the other hand, are small, reliable, and energy-efficient, which made it possible to build faster, more powerful computers. Early computers—used primarily for military and scientific calculations—were large, expensive, and limited in capability. However, the transistor ushered in a new era of miniaturization and efficiency, enabling rapid advancements that transformed computing forever.

2−1−2

As transistor technology improved, engineers were able to fit more transistors onto a single microchip, a process known as exponential transistor scaling. By the 1960s, this innovation allowed chips to integrate hundreds of transistors, laying the foundation for the modern computer era. As transistor density increased, it enabled greater processing power, expanded data storage, and enabled a reduction in the size and cost of computers. This breakthrough sparked a technological revolution that impacted nearly every industry—from telecommunications to healthcare —and transformed daily life. The key to this progress was the continual shrinking of transistor size, which allowed engineers to fit more onto a chip with each new generation, directly boosting computing power. By the 1980s and 1990s, the effects of this scaling were evident as personal computers became a common household and workplace technology, the Internet connected the world, and data-driven applications reshaped industries and society at large. This period is often seen as the Golden Age of Computing, defined by rapid advancements driven by the exponential growth of transistor technology.

2—1—3

Amidst all this growth and promise, Gordon Moore, co-founder of Intel Corporation, made a key observation in 1965: the total number of transistors able to be placed on a single microchip would double about every two years, which would result in an exponential growth in computing power. This observation, now known as *Moore's Law*, was met with immense optimism. For decades, it served as a guiding principle, with engineers and manufacturers striving to meet its ambitious predictions. The belief was that computing power would continue to grow at an unprecedented rate, unlocking new possibilities that seemed limitless. The industry embraced

Moore's Law as its roadmap, with each new chip release signaling the potential for continued progress. However, as time passed, it became evident that *Moore's Law* had limitations. The relentless push for miniaturization began to bump up against the fundamental limits of material properties and physics. What had initially seemed like a law of nature became increasingly difficult to sustain, and the excitement surrounding exponential progress started to wane. The once bright future foretold by *Moore's Law* gave way to caution, as the industry began to ask not "How quickly can we scale?" but "How long can we keep scaling?" The inevitable plateau in transistor scaling seemed closer with each passing year.

Section 2 — The Plateau Begins: The Slowdown of Progress Promised by Moore's Law and Its Impact on Progress

2−2−1

By the early 2000s, the industry began to recognize that the exponential progress predicted by *Moore's Law* might not be sustainable much longer. What had once been a predictable march toward smaller, faster, and more powerful chips started to show signs of slowing down. The relentless drive to shrink transistors and pack more of them onto a single chip ran into the physical limits of materials and the fundamental laws of physics. As transistors approached sizes of just a few nanometers, new challenges emerged. Issues like heat dissipation, quantum effects, and the physical constraints of silicon itself became increasingly difficult to overcome. At first, the slowdown was subtle, often masked by clever innovations in architecture, such as multi-core processors and parallel computing. Manufacturers also sought to maintain performance gains by improving chip designs and leveraging new technologies like 3D stacking—a method of layering multiple semiconductor layers to improve speed and efficiency—and experimenting with alternative materials such as gallium arsenide and silicon-germanium, which offered potential advantages over traditional silicon. However, these advancements, while impressive, could no longer match the pace of progress outlined by *Moore's Law*. The grand expectation that computing power would double every two years gradually began to feel like a distant memory.

2—2—2

As the 2010s unfolded, the pace of transistor scaling continued to slow, and the broader impact on technological progress became increasingly evident. Industries that had once relied on constant, rapid advancements in computing power began to experience a plateau in innovation. For example, fields like big data and high-performance computing, where large-scale processing power had been a driving force, faced significant challenges due to diminishing returns in line with the outlook previously forecasted by *Moore's Law*. New breakthroughs were no longer occurring at the same accelerated rate, and companies had to rethink their strategies for improving performance. This shift marked a fundamental turning point. As the once-unquestioned assumption of perpetual exponential growth in computing power came under scrutiny, the focus shifted toward alternative avenues. Quantum computing, still largely theoretical, held the promise of massive leaps in computational power, but the technical and practical barriers remained immense. Specialized processors, such as GPUs designed for specific workloads, offered incremental improvements, but these too could only push the envelope so far without a true breakthrough in hardware. Similarly, software optimization became a key focus, with increasing emphasis placed on finding efficiencies in code and workflows. However, this was never intended to be a replacement for the raw, exponential power gains that *Moore's Law* had once touted.

2—2—3

As the industry grappled with the stagnation caused by the plateau, it became clear that the era of exponential growth in computing power was drawing to a close. The once-dominant mantra of "bigger, faster, cheaper" no longer seemed to promise the same level of progress, and with the traditional route of transistor scaling facing insurmountable challenges, the tech world found itself confronting a critical juncture. Companies and engineers, accustomed to constant leaps forward, now found themselves scrambling to identify the next breakthrough to propel the industry forward. This uncertainty sparked a frantic search across Silicon Valley, where businesses explored every possible avenue—from quantum computing to novel hardware architectures, and even alternative software paradigms—in a race to uncover the next transformative technology that could drive progress. The quest for innovation had shifted from being merely a pursuit of progress to becoming a matter of industry survival.

Section 3 — Silicon Valley's Existential Crisis: Companies Scrambling for the "Next Big Thing"

2—3—1

As the explosive growth in the computing world began to slow, the once high-speed tech industry found itself at a crossroads. Silicon Valley, long synonymous with boundless growth and innovation, was now facing the reality of a future that no longer promised the same relentless acceleration. This shift brought about an existential crisis for the technology capital. The inability to sustain the advances in computing power predicted by *Moore's Law*, coupled with the physical limits of silicon, left companies scrambling for a new frontier. The belief that exponential gains in processing power would continue indefinitely was shattered, and with it, the industry's vision for the future. Tech giants and startups alike poured resources into every emerging technology, hoping to find the one that could reignite the flame of progress. In this race for the "next big thing," quantum computing emerged as a central focus. However, despite its vast potential, quantum computing remained in the realm of theoretical research, requiring enormous investments of time and capital. Meanwhile, established players like Intel, AMD, and NVIDIA shifted focus toward specialized processors, such as GPUs, that could handle specific workloads more efficiently. While these incremental advances were useful, they were hardly the breakthrough the industry desperately sought. At the same time, alternative architectures and entirely new computing models began to surface, each promising to overcome the stagnation. Yet, none offered a clear path forward that could rival the dramatic leaps of the past. Amid the uncertainty, a deeper crisis emerged: it wasn't just about maintaining the pace of progress —it was about securing the future of the industry itself. Companies that once relied on transistor scaling for innovation were now searching for new paradigms to redefine computing altogether—its potential and its direction. The pressure was immense, and the stakes were high: without a breakthrough, the dominance that Silicon Valley had enjoyed for decades could quickly fade into extinction. The industry's obsession with the next big thing was no longer simply about innovation—it had become a fight for survival.

2—3—2

As the industry raced to secure its future, the quest for a game-changing breakthrough became a frantic scramble. Since quantum computing was still largely theoretical and specialized processors offered only incremental improvements, Silicon Valley's technological leaders were forced to entirely accept the harsh reality that their previous model of explosive growth no longer applied—it was no longer just a cautionary indicator but an imminent threat. The focus began shifting toward entirely new frontiers—whether through alternative computing architectures, rethinking software design, or exploring the potential of biologically inspired systems. The very definition of innovation was being redefined, as companies shifted their attention from merely hardware to reshaping the foundational assumptions about what computing could achieve. In this environment, failure to adapt could mark the end of an era, and the once-celebrated ideals of Silicon Valley were now at stake. Yet, amid this uncertainty, the industry began to realize that its survival might not lie in finding the next big leap, but in learning how to sustain marginal progress in a world where exponential gains were no longer guaranteed.

2—3—3

In the face of these challenges, Silicon Valley's search for a new direction took an unexpected turn, as the desperate scramble for innovation shifted to rebranding the industry itself. With the promise of limitless growth through hardware advances debunked, the concept of "Artificial Intelligence" began to reemerge as the perfect solution to fill the growing void. With its broad, yet vague potential, "AI" offered a new lens through which the technology world could reframe its narrative of progress. Although "AI" remained fundamentally just a tool—limited by the constraints of its design —it was increasingly marketed as something far more profound: a symbol of innovation, the next great frontier that could supposedly propel the industry forward. This shift wasn't driven by groundbreaking breakthroughs in the field, but by the strategic repackaging of existing technologies and ideas under the "AI" umbrella. As a result, companies that had once focused on transistor scaling and hardware breakthroughs now began to tout "AI" as the new paradigm. What had once been a niche research area suddenly became the centerpiece of Silicon Valley's narrative, allowing the industry to project an image of continued forward momentum—even as the pace of true technological breakthroughs had slowed. In this sense, "AI" became not just a technology, but a lifeline for an industry trying to convince itself—and the world—that the future of progress was still within reach.

Section 4 — Desperate for A New Narrative: How Redefining "AI" Conveniently Filled the Innovation Gap

2—4—1

The technology industry, grappling with its technological stagnation, found a convenient solution in the concept of "AI." This idea emerged as a critical response to a pressing problem: how to maintain momentum in a world where the promise of exponential growth had undeniably faltered. "AI," in its broadest sense, offered a way to create a new narrative for the technology industry, one that could recapture the optimism that had defined the tech boom of previous decades. While "AI" itself was not entirely new, it had advanced in its perceived sophistication and gained widespread attention, making it the perfect vehicle for a rebranding effort. What had once been a niche field of academic research—focused on applications like speech recognition, image processing, and early forms of "machine learning"—had now expanded to encompass a broad spectrum of technologies and aspirations. "AI" was magnified to become a catch-all term that conveyed the illusion of limitless potential, allowing the industry to present itself as still on the cutting edge, even in the absence of the hardware breakthroughs that had once driven its progress. The timing of this rebranding effort was crucial. As Silicon Valley struggled to push the boundaries of silicon-based computing, the "AI" narrative enabled companies to promise a new era of progress without needing immediate, groundbreaking advancements. By framing the future of computing as a shift toward "intelligent" systems, companies could claim they were still pushing the boundaries of innovation, even if the core technology itself hadn't fundamentally changed. Instead of focusing on the limitations of hardware or the slowing pace of advancement stated in *Moore's Law*, companies marketed their products and innovations as part of the "AI revolution," promising everything from "smarter" algorithms to "self-learning" systems that could transform industries. In a world that had come to expect rapid, continuous advancements, the idea of "AI" provided a reassuring narrative of progress, helping to mask the stagnation in the underlying technology.

2—4—2

The concept of "AI," however, was not without its contradictions. While the rebranding of "AI" helped the industry maintain an image of cutting-edge innovation, it often glossed over the fact that many of the technologies marketed as "Artificial Intelligence" were far from revolutionary. In many instances, these were merely the latest iterations of established technologies, such as "machine learning" algorithms, data-driven optimization techniques, and statistical models—packaged and marketed with trendy new labels. Despite the overhyped expectations surrounding it, "AI" rapidly became a central talking point in nearly every company's strategy, from hardware giants to software startups. Even established companies like Google, Microsoft, and IBM significantly ramped up their "AI" initiatives, shifting their messaging and marketing strategies to capitalize on the emerging trend. The focus wasn't necessarily on what "AI" actually was, but rather on what the public could be made to believe "AI" represented: the next great leap in technology, the saving grace for an industry that had reached a plateau. In this sense, "AI" quickly became more of a branding tool than a concrete technological breakthrough. Yet, in a world desperate for a narrative of progress, this rebranding effort was more than enough to fuel the industry's need for a "next big thing."

2—4—3

As if the technological stagnation wasn't enough, the outbreak of the China Virus (COVID-19) pandemic added a new layer of complexity to the industry's already fragile situation. The pandemic forced businesses into crisis mode, rapidly shifting workforces to remote setups, halting production, and creating widespread economic uncertainty. With in-person collaboration disrupted and global supply chains buckling under strain, tech companies found themselves struggling to maintain what momentum they had left. The disruptions brought by the pandemic deepened the sense of urgency for a new wave of innovation that could help stabilize the industry. In this context, "AI" offered a dual promise: not only was it positioned as the future of technology, but it also became an ideal solution for navigating the challenges brought about by the pandemic. While technologies like process automation, remote systems, and data-driven adaptation had been in use for years, the rebranding of these tools as "AI" gave them the appearance of a cutting-edge transformation. Suddenly, the established tools of the tech world— the low-hanging fruit—were marketed as vital components of an "AI-driven" future, creating the illusion of rapid, revolutionary change. As the world grappled with an uncertain future, "AI" became more than just a marketing term; it transfigured into a symbolic lifeline, offering hope for the industry's recovery and, perhaps, even the transformation of society at large. In many ways, the pandemic acted as a catalyst, intensifying the race to rebrand and redefine what the future of technology could look like, further solidifying "AI's" narrative of potential—even if much of it was based on repackaged existing technologies.

"AI" Means Fraud

"Fallacies do not cease to be fallacies because they become fashions."

G.K. Chesterton
Illustrated London News, 1930

"There are two ways to be fooled. One is to believe what isn't true; the other is to refuse to believe what is true."

Søren Kierkegaard
Works of Love, 1847

Chapter 3:
Defining the Deception (What "AI" Actually Is—and What It Isn't)

Section 1 — Automation, Algorithms, and "Machine Learning": Breaking Down What These Technologies Actually Do

3—1—1

Having navigated the significant milestones in the "AI" timeline, from its early conception to its present-day positioning as the industry's buzzword du jour, we've now arrived at a critical juncture. We've examined the origins of "AI" and traced its progress from hope-filled concept to disappointment-dripping narrative, so next we'll dissect that fraudulent narrative. The so-called "AI revolution" has been marketed as the dawn of a new era, but when we peel back the layers of hype, marketing jargon, and outright deception, we find that the reality is far less impressive than the claims. What is "AI," really? What do these systems actually do, and what makes them fundamentally different from the technologies we've been using for decades? We've seen the term "AI" applied liberally to a wide range of technologies, from simple automation systems to sophisticated "machine learning" algorithms. However, at its core, what is being marketed as "AI" today is built on familiar concepts that are far from new, revolutionary, or "intelligent." To understand the deception, we need to break down the key components that have been falsely conflated under the umbrella of "AI," such as automation, algorithms, and "machine learning," and examine exactly what they are and what they are not. The effectiveness of the lies propelling the "AI" fraud resides in their ability to blur the lines between the illusion of intelligence—crafted using well-understood computing principles—and actual "Artificial Intelligence". By conflating automation, algorithms, and "machine learning" under the vague, misleading umbrella of "AI," the tech industry has created a false sense of progress and mystique. While these technologies are useful in their own right, they are not new, nor are they intelligent. Let's break them down individually and examine what they actually do.

3−1−2

Automation is one of the oldest and most fundamental applications of computing, predating the modern "AI" hysteria. At its core, automation involves the execution of predefined, rule-based tasks with minimal human intervention. Whether it's an assembly line robot welding car parts, a bank system processing scheduled transactions, or a script sorting emails into folders, automation is simply software controlling machinery or systems, following pre-programmed instructions. Unlike intelligence, which requires reasoning and undefined adaptability, automation operates within a fixed set of parameters defined by human programmers. Granted, modern automation has progressed to become more flexible than its predecessors. Today's systems can incorporate more complex rule sets, integrate with various large data sources, and dynamically adjust workflows based on real-time inputs. However, this added sophistication does not equate to intelligence. For example, an automated fraud detection system might flag transactions based on a broader range of variables than an older system, but it is still following predefined conditions—it does not "think," "understand," or make independent decisions. Despite this, the tech industry has eagerly rebranded automation as "AI" to capitalize on the general public's lack of understanding of what it actually is. Tasks that were once acknowledged as simple, rule-based processes are now repackaged with misleading labels to create the illusion of innovation. Business automation tools, which have existed for decades, are suddenly promoted as "AI-driven workflow solutions." Automated customer service systems that once simply followed structured decision trees are now marketed as "AI-powered virtual assistants." Even software that automates data entry, report generation, and email filtering is falsely touted as an "AI" breakthrough, despite relying on the same fundamental logic it always has. This fraudulent rebranding works because most people assume that if something is labeled as "AI," it must be fundamentally more advanced than its predecessor—specifically, that it now possesses the ability to reason and make independent decisions. In reality, nothing has changed except the terminology. While automation has become more adaptable, it remains exactly what it has always been: predefined logic executing tasks without comprehension, awareness, or intelligence.

3—1—3

Now that automation has been covered, the next logical topic to address is **algorithms**. The term "algorithm" has been thrown around as though it carries some kind of mystical power, but in reality, an algorithm is simply a structured set of instructions designed to accomplish a specific task. Algorithms are the very foundation of automation —what enables software and systems to process inputs, sort data, and execute commands in a structured way. However, despite their crucial role, algorithms do not possess intelligence, awareness, the ability to "think," or the ability to "learn." From the simple arithmetic steps in a calculator to the complex sorting mechanisms behind search engines, algorithms are the fundamental building blocks of computing, yet they have existed for centuries—the Euclidean algorithm dates back to 300 BC—long before the advent of modern computers, and certainly long before the rise of their fraudulent "AI" branding. Despite this historical context, the tech industry has deliberately blurred the line between algorithms and intelligence. By attaching the vague label of "AI" to systems that use algorithms, companies suggest that these systems are capable of "thinking" or making autonomous decisions. In reality, the algorithm itself is just a set of predefined rules—it cannot learn or reason. It cannot even adapt unless explicitly designed to do so through additional layers, which require the involvement of human programming. The industry's deceptive conflation of algorithms with "AI" relies, again, on the fact that most people do not understand what an algorithm actually is. This lack of understanding allows companies to sell mundane computational processes as something revolutionary. If an online platform uses an algorithm to recommend products based on a user's past behavior, that's just pattern matching—there's no "thinking" or "intelligence" involved. If a credit card fraud detection system flags a suspicious transaction, it's simply running predefined statistical checks—no "learning" or autonomous decision-making is taking place. Even advanced search engines, which rapidly sift through massive datasets, are still executing carefully crafted procedures designed by human engineers. The reality is that an algorithm—no matter how sophisticated—remains a tool that follows instructions. The industry's push to conflate algorithms with "AI" is as absurd as calling a vending machine "intelligent" because it dispenses a snack when the correct amount of money is inserted. Just because a system can process inputs and generate outputs based on predefined logic does not mean it is "thinking" or "understanding" anything. Yet, by dressing up these basic computing principles with deceptive language, companies have managed to convince the world that the same algorithms they've been using for decades are now part of some grand technological revolution. In truth, the only thing that has changed is the marketing.

3—1—4

Next, let us take a look at "**machine learning**." Perhaps the most strategically deceptive term in the entire "AI" lexicon, "machine learning" is a phrase designed to invoke the idea of human-like cognition—suggesting that machines can absorb information, develop understanding, and adapt in ways that resemble actual learning. This is a lie. "Machine learning" does not actually involve learning in any sense comparable to human learning—it is simply a statistical process that refines outputs based on patterns in data. There is no comprehension, no reasoning, and certainly no intelligence—just probabilistic guesswork dressed up as something far more profound. At its core, "machine learning" is essentially advanced curve-fitting. Here's how it works: a system is fed large volumes of historical data, from which it detects correlations and uses those patterns to make predictions. This is why a "machine learning" model can identify a face in a photo or suggest the next word in a sentence—not because it understands anything about faces, language, or context, but because it has been trained on vast datasets with similar examples. It applies statistical weights to inputs and produces outputs based on probability, not any form of understanding. And yet, the industry continues to insist on calling this process "learning," willfully exploiting the public's lack of understanding about technology to create the illusion that machines are developing knowledge and understanding. This deliberate misrepresentation is a critical pillar of the fraudulent "AI" narrative. The more people believe that machines are "learning," the easier it becomes to sell the fantasy that "AI" is something resembling true intelligence. However, strip away the deceptive—and often blatantly fraudulent— marketing, and these systems are nothing more than sophisticated statistical models—advanced in design, but fundamentally incapable of thought, reasoning, or autonomy. The only real intelligence at play is that of the people using misleading language to make statistical analysis appear to be something far more revolutionary than it actually is.

Finally, let's briefly focus on **Large Language Models** (LLMs). Rather than being a revolutionary leap in technology, LLMs are simply the latest—and most aggressive—iteration of statistical pattern-matching. Their defining trait is scale: trained on massive datasets, they generate text that appears knowledgeable, not because they understand language, but because probabilistic modeling at extreme scale creates an illusion of depth. The fluency of their output is a confidence trick—people mistake smooth, human-like responses for intelligence when, in reality, these systems are blindly remixing patterns without comprehension. In other words, just because an LLM can respond in a way that seems thoughtful or natural, that doesn't mean it actually understands what it's outputting—it's simply stringing together words based on statistical probabilities, with no real grasp of meaning behind them. A great comparison of LLMs are the so-called expert systems of the 1980s, which were marketed as demonstrating "AI's" ability to replicate human expertise. These systems were rigidly rule-based, relying on a fixed set of knowledge encoded by human experts to make decisions in narrow domains. Although, unlike LLMs, which can adapt to massive amounts of unstructured data, expert systems were rigid, following predefined paths set by their creators. That said, both expert systems and LLMs are rooted in the illusion that computers can reason or make decisions autonomously. The key difference lies in the methods: expert systems were grounded in rigid rule sets, whereas LLMs, though far more dynamic, are statistical engines that produce plausible-sounding text based on the data they process. Worse, though, an LLM's indiscriminate absorption of data means that they can replicate misinformation, biases, and errors as readily as facts, presenting falsehoods with the same confidence as truth. Even with continuous moderation and censorship by their human creators, LLMs still often retain vast amounts of erroneous and harmful data. Furthermore, the need for such oversight only further underscores the fundamental lack of intelligence and autonomy in these systems. As more LLM-generated content circulates, even under human supervision, these models will continue to adapt based on flawed data, perpetuating a cycle of diminished quality while reinforcing the illusion of intelligence. This is important to note, because it is precisely these statistical parlor tricks that falsely give popular chat solutions like Grok and ChatGPT their notoriety and appeal—convincing users that they are engaging with something intelligent rather than an advanced form of autocomplete. In the end, LLMs do not bring us closer to intelligence—they are simply a more polished, more deceptive repackaging of the same empty statistical processes that have long defined "AI."

Section 2 — The Rebranding of Old Tech as "AI": How Companies Slap the "AI" Label on Everything

3—2—1

The tech industry's obsession with the "Artificial Intelligence" narrative has led to an undeniably absurd trend: slapping the term "AI" on anything that even remotely resembles automation or data processing. This isn't an accident—it's a deliberate strategy designed to manipulate consumers and investors into believing that ordinary, even outdated, technologies are groundbreaking. The reality? These so-called "AI" technologies are nothing more than rebranded versions of old systems, often with incremental—though arguably useful—improvements. It's like adding a few extra gears to an old bicycle and calling it a "revolutionary new transportation device." The enhancements might make the system run a little faster or smoother, but they don't fundamentally change how it works. Take the so-called "AI" behind modern customer service chatbots. These have existed for decades, designed to handle simple customer inquiries using predefined responses structured in decision trees. They weren't "intelligent" back then, and they certainly aren't now. So why are companies aggressively marketing them as "AI-powered virtual assistants"? It's nothing more than a calculated marketing ploy—an effort to exploit the "Artificial Intelligence" hype and make something as mundane as a chatbot seem groundbreaking. Though, chatbots aren't the only examples. The same deceptive strategy is applied across a wide range of technologies. Predictive analytics, which has long been a fundamental tool in business, is now repackaged as "AI-driven insights." Voice assistants like Siri and Alexa are hailed as "AI-powered," when in reality they are just improved versions of earlier speech recognition and natural language processing systems—technologies that have existed for decades. The underlying mechanisms remain largely the same; the only real difference is that today's systems have access to larger datasets, slightly better statistical models, and improved integration with other software. That's it. It's not "AI." It's just incremental progress in automation and data processing, but by slapping an "AI" label on it, companies make it seem futuristic and sophisticated. And because I cannot hammer down this point enough, this is deliberate manipulation of language orchestrated to play on the general public's ignorance of how these technologies actually work. Yes, the tech leaders and marketeers behind this fraud think you're too incompetent to ever know any better. Don't worry, though, I'm going to help you understand what some of the most common "AI" tools or technologies you regularly come across actually are.

3—2—2

Chatbots—And since we've already begun to examine them, chatbots provide one of the clearest examples of how the "AI" label is being misused to deceive the public. These tools have existed for decades, initially designed to handle user inquiries with predefined scripts or decision trees. They weren't "intelligent" then, and they're still not "intelligent" now. Modern chatbots, like Grok and ChatGPT, use more advanced versions of the same basic principles, relying on natural language processing (NLP) and "machine learning" algorithms to match user input with the most relevant response. These systems can handle a broader range of queries and even automate tasks that once required human expertise. However, these tasks are things computers could already do. While some modern chatbots may appear more sophisticated, they still rely on data patterns—not understanding or reasoning. Chatbots don't comprehend context, they don't make independent decisions, and they certainly don't "learn" (they adapt—or at least try to). So why are they branded as "AI"? It's because the label makes them seem revolutionary, even though they're just statistical models built on the same principles as earlier systems. The only real difference is the marketing. By rebranding chatbots as "AI," the businesses selling them are able to sell an illusion—one that suggests these systems are capable of cognitive behavior that mimics that of a human. In reality, they're just advanced tools that return responses based on data patterns, offering no true intelligence. The growing use of this deceptive label does more harm than just misleading customers—it feeds the false narrative that true "AI" has finally arrived when, in fact, it hasn't and will likely never come to fruition (I'll explain why I believe it won't in the conclusion of this book).

3—2—3

Voice assistants—The next common misuse of the term "AI" is in the rebranding of voice assistants (think Siri and Alexa), which nowadays are marketed as being "powered by AI." In reality, they are just chatbots with a voice interface. Much like text-based chatbots, they rely on predefined algorithms to match input (in this case, spoken words) with relevant responses. They use speech recognition and natural language processing (NLP) to identify and interpret voice commands, but their apparent understanding is still just based on pattern-matching and statistical models—not true comprehension or reasoning. The only real difference is the medium of interaction: chatbots use written words, while voice assistants use spoken words. However, it seems like modern iterations of these are now utilizing both mediums. Despite being branded as "AI," these voice assistants do not understand the nuance of spoken language, including tone, emotion, or inflection. While they can recognize basic inflective patterns in speech—identifying a question versus a command —this doesn't equate to true understanding or a genuine mastery of human vocal inflection. They don't "hear" a conversation in the way a human does, nor do they process the emotional weight behind words. In reality, these systems are far from exhibiting the kind of intelligence or understanding that the term "AI" implies. Instead, they are simply sophisticated, voice-driven chatbots that are highly optimized for specific functions, but still fall short of achieving anything resembling human-like understanding or reasoning.

3—2—4

Image generation—Next, let's take a look at the so-called "AI" image generation tools, like DALL·E, MidJourney, and a myriad of others, all of which are marketed as revolutionary advancements in "Artificial Intelligence." The claim is that these tools can generate images from natural language descriptions, and while they can often produce visually impressive results, these tools are not "intelligent" in any sense. They do not create anything in the way a human artist does, nor do they possess any awareness, understanding, or conceptual thought. These systems rely on algorithms trained on massive datasets of images and their corresponding textual descriptions. When given a prompt, these systems don't produce an entirely new image through independent artistic thought. Instead, they generate a novel arrangement of pixels based on statistical patterns extracted from their training data. While this process can result in visually striking and seemingly creative outputs, there is no underlying artistic intent, comprehension, or originality—only pattern recognition and probabilistic modeling. These systems aren't "thinking" or "creating"; they're assembling outputs based on mathematical extrapolation of previously learned data. In other words, it's like how modern pop songs are all built from the same formula—catchy hooks, predictable beats, and a little auto-tune magic. Sure, it might make you tap your foot, but it's not the result of creative genius, it's just the probability of what's likely to work based on what's worked before. The "hit" is essentially generated by finding the most statistically probable combination of sounds, just like image tools generating pictures based on patterns, not inspiration. Calling these tools "AI" fuels the false belief that they possess cognitive abilities, like human creativity. In reality, they are nothing more than high-powered statistical models of visual patterns, repackaged as something more profound than they actually are.

3—2—5

Autonomous vehicles—Abstracting from our examples linked to typical consumer computing tasks, let's next consider "autonomous vehicles," such as those marked by companies like Tesla, Waymo, and others. While they are aggressively marketed as "AI-powered," this label is outright deceptive—especially when the technology behind them doesn't even approach true autonomy. These vehicles leverage a combination of sensors, cameras, radar, and predefined algorithms to process their surroundings and execute driving functions. However, they don't "understand" their environment in any sense. They do not perceive the road, anticipate events, or make decisions as a human driver would. Their actions are entirely dictated by statistical models that match input data to programmed responses. Despite being labeled "autonomous," these vehicles do not make independent decisions; they execute preprogrammed reactions to patterns in the data they collect. They are incapable of adaptation to a degree that's comparable to the skill and reaction time of a competent, trained human driver and cannot handle unforeseen situations without additional human intervention or software updates. Calling these systems "autonomous" is not just misleading—it's flat-out wrong. The reality is that these vehicles are complex, highly-automated machines following preprogrammed instructions, not thinking entities making autonomous choices. The fraudulent branding of these vehicles is yet another example of marketing-driven deception, reinforcing the illusion that true "Artificial Intelligence" exists when, in fact, we are still dealing with nothing more than advanced automation.

3—2—6

As you should now fully grasp, hopefully, the term "Artificial Intelligence" has become little more than a marketing buzzword—applied deceptively to technologies that, while sometimes impressive, fall woefully short of embodying true intelligence. Whether it's chatbots, voice assistants, "autonomous vehicles," or image generation tools, these so-called "AI" systems are nothing but advanced tools built on predefined algorithms, statistical models, and pattern recognition. They don't have comprehension, reasoning, or creativity —hallmarks of true intelligence. Instead, they simply replicate tasks based on enormous datasets. The use of terms like "AI" or "machine intelligence" give the false impression of cognitive abilities that these systems don't actually have, fueling widespread misconceptions about what they can actually do. And as if I haven't stated this enough before, this hype isn't driven by groundbreaking innovation but by savvy marketing designed to exploit the public's fascination with "AI" and its general lack of understanding of the actual technology behind it. In the next chapter, we'll delve into how this deceptive narrative serves certain industries and explore the question: who profits from the lies?

Section 3 — What "AI" Adds Is More Data, Bigger Processing Power, and Slicker Packaging

3—3—1

Behind the buzzwords and flashy marketing, what's called "AI" is far from groundbreaking—it's essentially just more data, greater processing power, and a sleeker package. More data simply means feeding vast amounts of information to statistical models, allowing them to make probabilistic predictions, not to "learn" or "understand." Greater processing power comes from scaling up existing infrastructure, using large numbers of conventional processors to handle bigger datasets more quickly, not the use of some new "AI-specific" hardware. A sleeker package refers to the user-friendly interfaces and smooth interactions, which mask the fact that these systems are resource-heavy and require immense computational power, often vastly more than typical corporate server setups require. Theses systems, promoted as cutting-edge innovations are, at their core, performing the same tasks computers have always handled—only on a much larger scale, and, yes, while I realize that I have hammered this point over and over again, it does bear repeating, because it's the absolute truth of the matter. There's no two ways about it. The "machine learning" models implemented in to such systems do not represent a new way of thinking or problem-solving; they simply apply brute-force computation to larger datasets than ever before. The ability to process massive amounts of information allows these systems to generate increasingly convincing outputs, but this is not a sign of intelligence—just an expansion of the same pattern-matching tricks that have always been the driving force behind so-called "AI" technology. The more data they absorb, the more polished their responses appear, and this creates a more impressive illusion of improvement without any actual progress toward comprehension or reasoning.

3—3—2

Meanwhile, companies frame this expansion of data-crunching as a technological leap, capitalizing on the public's limited understanding of computing to sell old ideas as cutting-edge innovations—yes, I am reminding you of this fact for the umpteenth time too. The fundamental mechanics of these systems remain unchanged, but their outputs look more sophisticated, their interfaces more seamless, and their branding more impressive. This is not the dawn of a new era in technology—it's a repackaging of statistical processes, dressed up to appear like something transformative. Ultimately, what "AI" truly represents is not an advancement in technology, but an inflation of scale—more data and more processing, paired with a carefully manufactured illusion of progress.

Section 4 — Why This Repackaging Is Wasteful and Brings About Unnecessary Sprawl: How So-called "AI" Technology Is Contributing to Inefficiency Rather Than Innovation

3—4—1

The sheer scale of the infrastructure that powers many of these so-called "Artificial Intelligence" systems is not just wasteful—it's actively detrimental to efficiency and sustainability. To perpetuate the illusions output by this deceptive technology, massive data centers are being constructed, consuming enormous amounts of energy and other resources to perform tasks that are fundamentally no more complex than traditional statistical operations. Such tasks could actually easily be handled by a personal computer from the 1990s. These centers, often marketed as the backbone of the future, require vast amounts of electricity, cooling, and physical space simply to execute what are, at best, trivial technological tricks. The reason these data centers consume so much is tied directly to the scale at which these systems operate. In order to achieve a level of "intelligence" that appears natural to the average user, these systems are designed to handle vast amounts of data across numerous models, running calculations at extraordinarily high speeds. The illusion of fluid, human-like interaction in these systems requires an immense amount of computational power—far beyond what traditional statistical models ever needed. For instance, generating a simple, coherent sentence in response to a prompt requires parsing through massive data sets and processing them in real-time. The scale of these tasks, especially when trying to simulate natural language, demands a corresponding scale in computational resources. Recall the recurring flaw in the history of "Artificial Intelligence"—the assumption that simply scaling up underperforming systems would eventually lead to genuine "AI." In reality, all this approach does is expand mediocrity, creating larger, more resource-hungry systems that still lack true innovation, merely amplifying the inefficiencies of their predecessors. This is the logically flawed pattern that is still being followed.

3—4—2

Additionally, it's worth noting that the widespread usage of these systems creates a feedback loop that exacerbates the situation. As more users interact with these systems, increasing their demand, the infrastructure needed to maintain the same level of performance also increases. More servers, more data processing power, and more electricity are required to meet the increasing demands of millions of simultaneous users. What was once manageable on a smaller scale quickly becomes a colossal energy sink as the system grows to handle a world of "always-on" services. This escalation, while marketed as an innovation, only highlights the excessive energy consumption and resource use necessary to maintain the illusion of "Artificial Intelligence," making the entire enterprise far more resource-intensive than the underlying tasks truly justify. And this cycle is ongoing—as long as the fraudulent "AI" narrative is perpetuated, the demand for resources will continue to climb. Of course, this serves the interests of those profiting from the scheme, as it ensures job security and sustained wealth. However, for the rest of us, it means that the needless strain on the electrical grid and the overconsumption of other resources will persist indefinitely.

"It is difficult to get a man to understand something when his salary depends upon his not understanding it."

Upton Sinclair
I, Candidate for Governor: And How I Got Licked, 1935

"The conventional view serves to protect us from the painful job of thinking."

John Kenneth Galbraith
The Affluent Society, 1958

Chapter 4:
The Business of the Hype (Who Profits from Perpetuating the "AI" Lie?)

Section 1 — How Venture Capital and Corporate Marketing Fuel the "AI" Gold Rush: The Financial Incentives Behind the Deception

4—1—1

There's no denying that the increasing reach of the term "Artificial Intelligence" has become a business phenomenon, with financial interests driving the promotion of technologies that don't live up to the hype. In this chapter, I'll expose how venture capital, corporate marketing, and deceptive advertising have converged to fuel the "AI" gold rush. From inflated promises to misleading branding, the so-called "AI revolution" is more-so about financial gain than it is about technological progress. By uncovering the financial motivations behind the deception, I'll show how venture capitalists and large corporations profit from a narrative built on empty promises, creating a cycle of inflated expectations and inevitable disillusionment. We'll also explore why, I believe, the legal system largely ignores these fraudulent practices and how this pervasive marketing strategy has continued unchecked, with little to no accountability.

4—1—2

At the heart of the "AI" gold rush lies a powerful financial engine: venture capital and corporate marketing. These two forces are central drivers behind the widespread hype surrounding so-called "AI" technologies, pushing them into the mainstream and positioning them as the newest big thing in tech. Venture capitalists, always hunting for high-return investments, have poured billions into companies that claim to offer cutting-edge "AI" solutions—often with little more than vague promises and lofty goals. The objective is clear: capture market share, attract additional investment, and profit from the perceived limitless potential of "AI". Corporate marketing teams have seized this opportunity by packaging these technologies to appeal to both consumers and investors. By branding products as "AI-powered" or "driven by AI," companies exploit the allure of innovation and intelligence, regardless of their actual technical capabilities. This marketing strategy not only stokes consumer demand, it also inflates the perceived value of the technologies: driving up stock prices, investment interest, and corporate valuations. The result is a cycle where hype perpetuates itself, all in the pursuit of financial gain —while the true capabilities of these technologies remain largely unexamined and unverified by their investors.

Section 2 — Case Studies of Overhyped "AI" Companies That Failed to Deliver: Examples of Major Failures

4—2—1

In this section, we'll examine a series of case studies featuring companies and products that promised groundbreaking advancements driven by so-called "AI," only to fall short of their hyped-up claims. These failures span a range of industries, from healthcare and biotech to autonomous vehicles. What unites them is the unethical and deceptive use of the "AI" narrative to attract massive investments and generate hype—often without delivering any meaningful results. These case studies reveal a recurring pattern of inflated promises, followed by failure and excuses, exposing how the fraudulent marketing of "AI" inevitably leads to disappointment and disillusionment. Keep in mind that what follows is, by no means, an exhaustive list of failures.

4—2—2

IBM Watson—Once hailed as the future of "Artificial Intelligence," Watson by IBM stands as one of the most prominent examples of the gap between the hype surrounding "AI" and its real-world limitations. Initially unveiled in 2011 as a system capable of answering questions in natural language, Watson gained widespread attention after its victory on the popular game show *Jeopardy!*

This success, albeit in a controlled environment, fueled the belief that Watson could revolutionize industries such as healthcare, finance, and customer service. IBM's aggressive marketing positioned Watson as an intelligent, "AI-powered" tool capable of analyzing vast datasets, offering personalized medical diagnoses, and improving business processes. However, as Watson ventured beyond a game show set, it quickly became clear that the reality did not match the hype. Despite massive investment and numerous high-profile partnerships, Watson's performance in real-world applications was underwhelming, and in many cases, disastrous. In healthcare, for instance, Watson was meant to assist doctors by analyzing medical records to diagnose conditions and recommend treatments. Yet, it soon became evident that Watson struggled with seemingly basic tasks, misinterpreting data and offering incorrect recommendations. In some instances, Watson even suggested unsafe treatments, putting patients' health at risk. Watson's failures in healthcare exemplified a broader trend of disappointing results across other industries. The system's inability to handle the complexities of real-world data—particularly in a critical field like healthcare—highlighted that Watson, while advanced in certain aspects of natural language processing, was far from the transformative "AI" technology that IBM had promised. Instead of the all-knowing, "autonomous" system that was marketed, Watson proved to be ill-equipped for the practical challenges of the industries it was supposed to revolutionize. Despite these failures, IBM did not abandon Watson. Instead, they rebranded and repackaged the technology in an attempt to salvage its image. As Watson's shortcomings became even more apparent, IBM pivoted its messaging, shifting from the promise of a groundbreaking "AI" technology for healthcare and other sectors to focusing on "AI-driven" "cloud" solutions. This rebranding allowed IBM to continue promoting Watson as a cutting-edge product without addressing its fundamental flaws. Rather than admitting defeat, IBM pressed on, continuing to market Watson's capabilities despite the lack of tangible, verifiable success in its initial applications. In the end, Watson became a symbol of the inflated promises made by companies marketing "AI" as a cure-all. IBM's failure to deliver on its promises, despite years of development and millions of dollars in funding, serves as a cautionary tale about the dangers of overhyping technologies. Instead of acknowledging its limitations, IBM's strategy of rebranding and repackaging Watson reveals how companies can perpetuate the "AI" myth, riding the wave of hype to continue profiting from a product that failed to live up to its grand promises.

4—2—3

Olive AI—Founded in 2012 by Sean Lane, Olive AI positioned itself as a revolutionary force in healthcare, claiming that its "AI-powered" platform would automate administrative tasks, optimize hospital workflows, and dramatically cut healthcare costs. The company aggressively marketed itself as the future of healthcare operations, boasting about its so-called "machine learning" and "artificial intelligence" technologies that were supposedly capable of handling everything from insurance processing to supply chain management. Olive raised over $800 million in venture funding based almost entirely on these promises, securing contracts with hundreds of hospitals and healthcare systems across the United States. Its valuation peaked at $4 billion, cementing its status as one of the most hyped "AI" startups in the healthcare sector. However, the underlying technology was never what the company claimed it to be. Former employees, industry insiders, and investigative reports revealed that Olive's platform was heavily reliant on basic automation scripts, manual human intervention, and rudimentary data processing—none of which even approached the level of true "Artificial Intelligence" that Olive loudly advertised. In many cases, the company's so-called "AI" simply amounted to offshore workers performing tasks behind the scenes, giving clients the illusion of intelligent automation.

Despite these glaring issues, Olive doubled down on its marketing narrative, portraying itself as an indispensable, "AI-powered" solution to healthcare's inefficiencies. By 2022, the facade had collapsed. After burning through hundreds of millions of dollars with little to show for it, Olive laid off a significant portion of its workforce and sold off large portions of the company at fire-sale prices. Venture backers lost most of their investments, and healthcare providers who had integrated Olive's solutions were left scrambling to replace critical workflows that had never truly been automated in the first place. Olive AI's collapse highlights how healthcare—one of society's most critical sectors—became a prime target for fraudulent "AI" narratives. By exploiting urgent institutional needs and masking rudimentary technology with buzzwords, Olive exemplified the dangers of allowing hype to substitute for real innovation, particularly where public health and financial stewardship are concerned.

4—2—4

Jawbone—Once a leading player in the wearable tech market, Jawbone serves as yet another example of a company that capitalized on the "AI" narrative to fuel its rise, only to collapse when its promises didn't materialize. Founded in 1999 by Alexander Asseily and Hosain Rahman, Jawbone initially gained recognition for its

Bluetooth headsets. However, it was the company's pivot into the wearable fitness tracker market in the early 2010s that truly captured the public's attention. Jawbone introduced the UP fitness band, which it marketed as a revolutionary product capable of tracking steps, monitoring sleep, and offering personalized health advice. Central to these claims was the idea that the device utilized advanced "AI" and "machine learning" algorithms to provide insightful data on users' health. At its peak, Jawbone was considered a strong competitor to Fitbit and other wearable tech brands, raising over $900 million in venture capital funding. The company's marketing painted a picture of an innovative, "AI-driven" fitness platform that could, not only track your movements, but predict and improve your health with intelligent insights. This narrative, driven by the allure of sophisticated "AI" technology, attracted a large consumer base and significant investment. However, the reality of Jawbone's technology did not live up to the hype. The UP fitness bands faced numerous hardware issues, such as malfunctioning sensors and battery problems, which led to high return rates and negative reviews. The company's claims about "AI-driven" health insights were similarly criticized for being vague and unsubstantiated. The "machine learning" algorithms that Jawbone boasted about were far from revolutionary, and the health data they provided was often inaccurate and lacked the depth promised by the company's marketing. As the wearable tech industry became more competitive, disillusionment set in. It became evident that Jawbone's technology was not as advanced as it had been marketed to be. The "AI-powered" health guidance that Jawbone promised users failed to fully materialize, leaving many consumers disappointed. In 2016, the company was forced to pivot, abandoning its wearable product line and shifting to a business-to-business model. Jawbone attempted to sell off its remaining assets, marking the beginning of the end for its venture into the wearable "AI" space. The company's decline was marked by a long and messy bankruptcy process, further illustrating the unsustainable nature of its business model, which was built upon false promises. Jawbone's failure to deliver on its "AI-powered" promises highlights the danger of relying on inflated claims about technological sophistication to attract investment and market share. Despite once being a frontrunner in the wearable tech industry, Jawbone ultimately could not keep up with the competitive demands of the market or deliver on the false and overly-hyped "AI" claims that had fueled its rise. Jawbone's downfall serves as a cautionary tale for companies that seek to ride the coattails of the "AI" hype—albeit without the technological foundation to back up their claims. Jawbone's overblown promises about its "AI" capabilities were not only misleading but unsustainable, ultimately contributing to its collapse in a fiercely competitive sector.

4—2—5

Uber's Self-Driving Car—In 2018, Uber found itself at the center of a controversy in Tempe, Arizona that would make headlines for all the wrong reasons. While it was operating in "autonomous" mode, one of its "self-driving" cars hit and killed 49-year-old Elaine Herzberg, who was walking her bicycle across the road. This tragic event became the first recorded death linked to a "self-driving" vehicle in the United States, quickly exposing the undeniable shortcomings of Uber's "AI-powered" technology. The vehicle, a Volvo XC90 equipped with lidar, cameras, and sensors designed to detect pedestrians, cyclists, and other obstacles, was meant to showcase the potential of Uber's self-driving program. However, during a nighttime test run, the car failed to avoid hitting Herzberg. Although the car's sensors reportedly detected her, the system failed to make the necessary life-saving decisions, and even more tragically, the safety driver—who was supposed to monitor the system—was distracted at the time of the incident. Uber had marketed its self-driving program as a breakthrough in "autonomous" transportation, promising a safer, more efficient future where self-driving vehicles could navigate complex environments and reduce traffic-related fatalities. Yet, the tragic accident highlighted the glaring gap between Uber's marketing and the reality of its technology. While the company claimed its vehicles were "AI-powered" and capable of navigating the road without human intervention, the incident revealed significant flaws in the technology's ability to function as promised. This failure underscored a broader issue: the frequent misuse of the "AI" label on technologies that lack the intelligence or capabilities their creators claim. In this case, Uber's self-driving car did not possess the critical decision-making capabilities needed to avoid hitting and killing Elaine Herzberg, despite the company's assertions that it was fully capable of handling complex driving scenarios. The incident served as a stark reminder of how "AI" is often marketed as a panacea, only for the reality of the technology to fall short of expectations. Uber's "self-driving" car failure is not an isolated example but part of a recurring pattern in which companies market so-called "AI" technologies as more advanced than they truly are. These technologies, when they don't outright deceive—which is a unicorn if there ever was one—simply fail to deliver on the promises made by their creators, as demonstrated by this fatal incident. If the proper application of existing laws and regulations had been in place, it's possible that this tragedy could have been avoided, and the public spared from the dangers of being misled into believing that "AI" could solve problems it was never designed to handle. Uber's failure to deliver on its "autonomous" vehicle promises stands as another cautionary tale in the broader discussion around the hype and deception surrounding "AI" technologies. The company's marketing of its "self-driving" car as an "AI-driven" solution was ultimately proven to be a tragic exaggeration of its true capabilities, further exposing the gap between hype and reality in the world of "Artificial Intelligence."

Section 3 — The Cycle of Inflated Expectations → Disillusionment → Repackaging: The History of Tech Hype Cycles

4—3—1

In the world of technology, a familiar pattern plays out: inflated expectations, followed by disillusionment, then a strategic repackaging of the same technologies to generate renewed interest and investment. This cycle has been especially persistent in the realm of so-called "AI." Each wave of hype brings grand promises—machines that will think like humans, revolutionize industries, and solve society's most complex problems. Yet time and time again, these promises fail to materialize. Rather than admitting fundamental shortcomings, companies rebrand and relaunch the same technologies under new narratives. This pattern dates back to the earliest days of "AI" research in the 1950s and 1960s. Pioneers like John McCarthy and Marvin Minsky were deeply optimistic, claiming that intelligent machines were just around the corner. Backed by institutions like IBM, Stanford, and MIT, early "AI" research carried the conviction that computers could soon rival human cognition. Yet, as the reality failed to meet expectations, disillusionment set in, funding dried up, and interest faded. Instead of abandoning "AI," however, researchers simply narrowed their focus —from general intelligence to specialized, narrowly defined tasks—marking the first full cycle of hype, disappointment, and rebranding that would come to define the field.

4—3—2

Thus far, each and every tech hype cycle has begun with an explosion of optimism. Companies, investors, and the media eagerly latch onto the next big thing, and "AI" has been no exception. The allure is obvious: smart machines that think, learn, and operate independently, promising to transform healthcare, finance, transportation, and more. The language surrounding "AI" tends toward grandiosity, with sweeping claims about products that will reshape daily life or even solve societal problems. Consider the frenzy around "autonomous" vehicles. Companies like Tesla,

Waymo, and Uber have made bold declarations about deploying fleets of self-driving cars, promising to reduce traffic accidents, lower emissions, and revolutionize mobility itself. Investors poured billions into startups, and media coverage amplified the narrative. Public excitement followed, fueled by the belief that these vehicles represented the inevitable future—all driven, supposedly, by "the power of AI."

4—3—3

Then, there's the crash that always follows the hype. Implementing so-called "AI" technologies in the real world has proven far more challenging than advertised. The autonomous vehicle industry, for instance, has been riddled with failures: high-profile accidents, the inability to handle unpredictable environments, and repeated public setbacks. The Uber fatality in 2018 and Waymo's struggles with real-world traffic are just the most visible examples. Other "AI"-branded ventures have suffered similarly. IBM Watson, once heralded as a breakthrough solution for numerous sectors, including healthcare, failed spectacularly—its cancer treatment recommendations were often inaccurate or outright wrong, leading to a major public backlash and the quiet dismantling of Watson Health. As these high-profile disappointments accumulate, investors begin to pull back, and media narratives turn negative.

However, instead of spelling the end, this stage usually marks the beginning of the next repackaging effort.

4—3—4

Rather than abandoning failed or underdelivering products, companies push the hype for as long as they can, adjusting narratives when scrutiny becomes unavoidable. Take Grok, for example. Initially marketed by Elon Musk as a revolutionary "AI" competitor to ChatGPT, it was ultimately repositioned as a next-generation search engine once the initial hype began to fizzle. "Grok it" is now the rallying cry—a tactical shift designed to salvage credibility without admitting failure. This kind of narrative adjustment is standard practice. Companies that once promised "Artificial General Intelligence" now claim they are building "specialized" models that "could lead" to general intelligence someday—if they get more time, more data, and, of course, more funding.

4—3—5

As I have pointed out repeatedly, this cycle is the hallmark of the tech industry's approach: companies inflate expectations, fail to deliver, and then reshape the narrative to keep the machine running. Silicon Valley has built an entire business model around it, and "AI" is simply the latest iteration of that model. As long as money can be made promoting overhyped products—and assuming companies aren't, more often, held legally accountable for deceptive marketing—the cycle will persist. The "AI" boom illustrates a broader reality about so-called breakthrough technologies: grand narratives often outweigh actual capabilities. While there are occasional incremental improvements, they rarely match the scale of the promises made. Ultimately, the persistence of the "AI" hype cycle reveals the true driving force behind much of modern technological development: not genuine transformation, but the relentless pursuit of narrative dominance and financial gain.

Section 4 — The Legal Aspect: Why Is This Fraud Seemingly Being Ignored?

4—4—1

In the U.S., several established laws aim to prevent fraudulent and deceptive marketing practices. The Federal Trade Commission (FTC) plays a central role in regulating and enforcing these laws. The *FTC Act*, for example, prohibits unfair or deceptive acts or practices in commerce, including misleading advertising and false claims. The *Lanham Act* likewise enables competitors to challenge false advertising in the marketplace. These regulations are intended to protect consumers from misleading claims about products and services. In the context of "AI" marketing, I feel compelled to conclude that many companies are blatantly violating these laws by making exaggerated, unsubstantiated, and often outright false claims about the capabilities of their products and services. Tesla's "self-driving" cars, for example, are publicly marketed as "AI-powered" despite clear evidence that they do not meet the standards of autonomous driving. Similarly, many "AI" tools, like ChatGPT, are sold with promises of delivering autonomous, self-improving intelligence, when in reality, they only simulate such behavior to varying degrees of believability. It seems, therefore, that the FTC should be stepping in to regulate these claims—yet such enforcement has been practically non-existent.

4—4—2

With the understanding that laws regulating fraudulent and deceptive marketing practices do, in fact, exist, I would like to assert why the clearly deceptive marketing of so-called "Artificial Intelligence" should be deemed illegal. I speak not as a legal expert, but as a concerned American citizen offering a logically reasoned argument. Under current regulations, companies are required to avoid misleading consumers with false or unsubstantiated claims. When businesses label products like xAI's Grok, Tesla's "self-driving" cars, or OpenAI's ChatGPT as "AI" solutions, they imply that these technologies possess the ability to make independent decisions, learn, and otherwise exhibit genuine traits of intelligence—whether on the human level or based on John McCarthy's definition of "Artificial Intelligence—none of which they can do. This is misleading at best and fraudulent at worst, creating false expectations and encouraging consumers and investors to make decisions based on misinformation. Fraudulent advertising is particularly dangerous in the tech sector, where hype around "AI" has become a goldmine for investors and consumers alike. Investors believe they are funding cutting-edge, world-changing technologies, while consumers purchase products they expect will radically improve their lives. When these promises fail to materialize, the harm—both financial and psychological—is very real. Take OpenAI, for example. ChatGPT is marketed as an "AI" technology that can engage in meaningful conversation, write essays, and assist with a myriad of tasks. Yet despite its arguable usefulness, it does not exhibit any true intelligence; it cannot think, reason, or learn, and it does not understand language or underlying concepts in the prompts it receives or the outputs it generates. Nevertheless, it is labeled "AI" to suggest capabilities far beyond its actual functionality. This is a textbook case of deceptive advertising—fraud by any other name—

except, it seems, when it comes to so-called "AI" technology.

4—4—3

The issue with "AI" marketing fraud is that it often goes unchecked by regulators, despite clear violations of deceptive advertising laws. I maintain that there are several reasons for this selective enforcement:

1. **Lack of Understanding:** The rapidly evolving nature of "AI" and the complexity of the underlying technologies often leave regulators struggling to keep pace. As a result, they may not fully grasp the implications of misleading "AI" claims or the extent of the deception they foster.

2. **Political and Economic Influence:** "AI" companies, particularly those led by figures like Sam Altman and Elon Musk, command enormous financial backing and political influence. With billions of dollars in funding and strategic partnerships, they are able to exert significant pressure on both the marketplace and regulatory bodies, contributing to a lack of accountability and reluctance to intervene.

3. **Big Tech Lobbying:** Major tech companies maintain strong lobbying efforts in DC, often shaping policies to their advantage. As "AI" becomes increasingly central to the economy, these companies lobby for lax regulations, arguing that strict oversight would stifle innovation—even though much of their so-called innovation rests on exaggerated or fraudulent claims.

4. **Inertia and Bureaucracy:** Even when regulators are aware of deceptive practices, significant delays often occur before any meaningful action is taken. The bureaucratic nature of governmental oversight, combined with limited resources and technical expertise, makes it difficult to address the complexities of modern "AI" narratives in a timely or effective manner.

4—4—4

That said, it seems evident to me that the lack of enforcement is not accidental. Several key players directly benefit from the continued proliferation of "AI" hype, despite its fraudulent nature:

1. **Tech Giants and Their Leaders:** Figures like Elon Musk, Sam Altman, and other

tech leaders profit immensely from "AI" hype. By keeping promises of groundbreaking "AI" alive, they attract massive investment. As companies like OpenAI and Tesla raise billions, these leaders expand their personal fortunes and influence. Even when the technology falls short, they continue to reap rewards—while shifting the consequences of failure onto consumers and investors.

2. **Investors:** Venture capitalists and institutional investors often focus more on "AI"'s speculative potential than its current capabilities. By sustaining the hype, they capitalize on rising company valuations, profiting from the promises of future breakthroughs, however exaggerated they may be.

3. **Government Officials and Regulators:** Some regulators may benefit indirectly through increased tax revenues, from political contributions, or even insider trading—an issue that has repeatedly surfaced among members of Congress. The rapid expansion of the "AI" industry bolsters local economies and political prospects, creating conflicts of interest that discourage enforcement.

4. **The Marketing and Media Machine:**
    Advertisers, media outlets, and PR firms
    all profit from the illusion of "AI"
    revolution. Tech companies spend
    billions promoting narratives of
    transformation. The media, in turn,
    thrives on sensational headlines, higher
    engagement, and lucrative advertising
    deals, creating a self-sustaining cycle of
    hype divorced from technological
    reality.

In my best estimation, the failure of regulators to intervene is not mere oversight—it is a natural outcome of financial and political systems that benefit from unchecked hype. Tech executives, investors, government figures, and media organizations all stand to gain from perpetuating the deception, at the direct expense of the public. Consumers are led to believe they are interacting with revolutionary machine-based intelligence when, in reality, they are paying for little more than glorified automation wrapped in deceptive branding. The fraud continues not because it is invisible, but because it serves those who profit from it most.

Section 5 — The Hypocrisy of Public Outrage: The "Vaccine" Controversy and the Silent Acceptance of Redefining "AI"

One of the most telling examples of public outrage over language manipulation in recent years occurred during the China Virus (COVID-19) pandemic. During that time, the term *vaccine* was redefined in a way that many perceived as a deliberate attempt to shape public perception. Traditionally, vaccines were understood to provide immunity—a concept deeply rooted in both medical literature and public consciousness. However, as the mRNA-based injections were rolled out, the definition of vaccine was quietly adjusted to mean a product offering some level of protection, rather than full immunity. This shift sparked intense backlash from those who saw it as an overt attempt to push a narrative under false pretenses. People rejected the idea of redefining a word to fit a political or corporate objective, recognizing it as a blatant manipulation of public trust. Yet when it comes to *"Artificial Intelligence,"* the same individuals who raged against the redefinition of *vaccine* remain eerily silent—even though the term "AI" has been hijacked in an identical fashion. Once meant to describe genuine machine intelligence—systems capable of thinking, reasoning, and improving independently—the term now refers largely to statistical models and predictive automation. Just as the redefinition of vaccine served political and corporate interests, the redefinition of "AI" benefits tech giants, investors, and governments seeking to control the narrative and profit from the illusion of revolutionary technology.

4—5—2

This silent acceptance of "AI" deception raises a critical question: Why was there mass resistance to the redefinition of *vaccine*, yet almost no pushback against the blatant redefinition of *"Artificial Intelligence"*? The answer lies in how language manipulation reshapes public perception. The technology industry, much like the pharmaceutical industry during the pandemic, relies on carefully curated narratives, media complicity, and public ignorance to protect its rebranded terminology from scrutiny. People resist redefinitions when they feel immediate impacts on their health or freedoms. But with "AI," the deception is more subtle— allowing mass manipulation to proceed largely unnoticed. Ultimately, the public's willingness to accept "Artificial Intelligence" hype, despite its obvious falsehoods, stems from a toxic mix of tech optimism, ignorance, and relentless media reinforcement. People *want* to believe that genuine "AI" has arrived, just as many wanted to believe in a quick medical fix for the pandemic. The media glorifies these technologies endlessly, drowning skepticism under a tide of corporate propaganda. In the end, the same mechanisms of deception that rebranded the meaning of *vaccine* are now rebranding *"Artificial Intelligence"*—and once again, the public is being played for fools.

"The glory which is built upon a lie soon becomes a most unpleasant incumbrance. ... How easy it is to make people believe a lie, and how hard it is to undo that work again!"

Mark Twain
Autobiographical dictation, 1906

"Every great cause begins as a movement, becomes a business, and eventually degenerates into a racket."

Eric Hoffer
The Temper of Our Time, 1967

Chapter 5:
The High Priests of the "AI" Hoax (The Key Players Behind the Fraud)

Section 1 — Attribution and Accountability: The Faces Behind the Fiction

5—1—1

Before diving into the profiles that follow, it's important to clarify the intent and scope of this chapter. The purpose here is to present a concise analysis of key individuals and entities who, in my professional opinion, have played central roles in the construction, promotion, and propagation of the modern fraudulent "AI" narrative. These are the architects, the salesmen, and the leading spokespersons—figures who have either knowingly shaped the narrative for strategic advantage or have recklessly amplified it without exercising the basic scrutiny that such claims demand, and without considering the negative downstream consequences. Keep in mind: the present "Artificial Intelligence" hype cycle isn't fraudulent because the technology does nothing useful or impressive, but because what the public has been led to believe about it is categorically false. It is not my aim to assign villainy arbitrarily, nor to dismiss genuine technological advancements. Rather, my goal is to shine a light on, in my reasoned opinion, some of the most prominent figures behind the narrative—because accountability begins with attribution.

The profiles that follow are not exhaustive, but they do represent, I believe, a cross-section of power, persuasion, and profit at the heart of the current fraudulent "AI" narrative. Let's briefly take a look at who made my list and why they made it. As you might have guessed, Elon Musk earns top billing as *The Contradictory Salesman*—a man who warns apocalyptically about "AI" while simultaneously profiting from its commercial application, most notably through Tesla's so-called "self-driving" technology. Sam Altman follows as *The OpenAI Puppet Master*, the man who helped morph a nonprofit research lab into a for-profit juggernaut fueling billion-dollar fantasies. Sundar Pichai appears as the architect behind *The Imitator General of Silicon Valley AI*, sparking a reckless rebranding spree throughout the tech giant's ecosystem. Satya Nadella, *The Mirage Investor*, emerges as a key figure aggressively embedding "AI" across Microsoft's empire to justify a $10 billion wager on OpenAI's promise. Then there's Mark Zuckerberg, *The Late Adopter Trying to Stay Relevant*, who pivoted from his failed Metaverse ambitions to seize a seat aboard the "AI" hype train. Rounding out the list are other key figures perpetuating the fraud: Jensen Huang of NVIDIA, who sells the hardware backbone of the dream; Demis Hassabis of DeepMind, who has built a career on overpromising human-level intelligence; and the tech media itself, whose uncritical recycling of corporate talking points ensures that the illusion remains both profitable and largely unchallenged. Each entry represents a sphere of influence capable of steering public perception and policy alike—and when examined together, what emerges is a consistent through-line of fear, hype, and opportunism wrapped in the deceptive language of progress.

Section 2 — Elon Musk: The Contradictory Salesman

5—2—1

Elon Musk has emerged as one of the most prominent figures on the "AI" scene, known for both his warnings about, what he frames as, the potential dangers of "Artificial Intelligence" and his simultaneous reliance on so-called "AI" technology in his businesses. Musk has repeatedly expressed concerns about "AI" being "the biggest existential threat" to humanity, a position he has emphasized in public appearances at venues like MIT, SXSW, and various interviews. He has advocated for regulatory oversight to prevent "AI" from becoming too powerful and uncontrollable. These cautionary statements have garnered him significant media attention, painting him as a reluctant, yet necessary, voice of reason in the tech world. However, Musk's actions often appear in conflict with his rhetoric. While warning of "AI's" dangers, Musk has actively promoted "AI-powered" technologies through his companies, most notably Tesla. Tesla's "self-driving" technology has been marketed as one of the most advanced applications of "AI," promising to revolutionize transportation and reduce road fatalities. Despite these claims, Tesla's Autopilot and "Full Self-Driving" (FSD) systems have faced substantial scrutiny due to a series of high-profile incidents, including fatal crashes involving the vehicle's supposed "AI-powered" capabilities. Musk continues to promote these technologies even as evidence accumulates that they fall short of the grand promises made in Tesla's marketing campaigns. Tesla's promotional materials often blur the lines between terms like "Autopilot," "self-driving," and "Full Self-Driving," creating the impression that Tesla's vehicles are capable of autonomy—despite clear limitations that require constant human supervision. Critics argue that this marketing strategy is not merely optimistic but deliberately misleading, presenting an image of autonomous capability that the technology demonstrably does not possess. Moreover, Musk has repeatedly predicted that Tesla vehicles would achieve actual autonomy—what he refers to as "full autonomy"— "next year," a promise, it seems to me, he has made every year since 2016, without fulfillment. Musk's contradictory stance—publicly warning about the dangers of "AI" while simultaneously profiting from its public image—exemplifies a broader strategy common in the tech sector: weaponize fear to boost credibility, then monetize inflated expectations. In Musk's case, the contradiction is not incidental; it is a deliberate and calculated formula, honed over years of grand promises, ambiguous language, and strategic marketing.

5—2—2

A glaring example of Musk's contradictory relationship with "AI" is his promotion of Grok, the so-called "AI assistant" integrated into the X social media platform. Despite Musk's frequent warnings about the existential risks of "AI," he continues to capitalize on the term's marketing power. The branding of Grok as a cutting-edge "AI assistant" is, at best, misleading. In reality, Grok is little more than a text-generation and automation platform dressed in natural language processing, offering a facade of intelligence that does not reflect genuine independent thought, learning, or self-improvement. Although Grok may leverage a large language model (LLM) backend—similar to other chatbot architectures—its functionality remains confined to predictive text generation, search aggregation, and basic task automation. It does not demonstrate any of the core attributes of "Artificial Intelligence," or even true autonomy. Its main distinguishing feature is its use of an edgy, sometimes vulgar, communication style—a superficial novelty in a saturated market of automated assistants. This edginess, far from signifying genuine intelligence, serves primarily as a marketing distraction. It gives users the impression of depth or sentience where none exists. By emphasizing provocative language, Grok masks its fundamental reliance on existing datasets, pattern recognition, and scripted interactions. In effect, Grok's "cutting-edge" reputation is not earned by technological innovation but manufactured through strategic presentation and deliberate exaggeration. Much like Tesla's "self-driving cars"—promoted as revolutionary but falling short of actual autonomy—Grok exploits the "AI" buzz to sell a product that ultimately rests on old principles of automation and prediction, not on any real breakthrough in intelligent systems. Musk's use of the "AI" label perpetuates the broader industry pattern: inflating expectations to drive hype, while delivering tools that are far removed from the transformative technologies they are marketed as representing. Through Grok, Musk has not only joined the "AI" hype cycle more directly but has also reinforced a business model built on grandiose promises and technological under-delivery.

Section 3 — Sam Altman: The OpenAI Puppet Master

Sam Altman, CEO of OpenAI, has played a central role in the company's transformation from a nonprofit with altruistic ambitions to a multibillion-dollar corporate powerhouse. Founded in 2015 to ensure the safe and ethical development of "Artificial General Intelligence" (AGI), OpenAI originally prioritized humanity's benefit over profit. However, its 2019 shift to a "capped-profit" model—bolstered by massive funding from corporate giants like Microsoft—marked a clear pivot toward financial gain. Under Altman's leadership, OpenAI has aggressively commercialized its technology, with products like ChatGPT at the center of a global tech race. ChatGPT's launch exemplified OpenAI's shift from ethical rhetoric to strategic monetization. Marketed as a revolutionary tool that could democratize knowledge and transform industries, its rollout was a masterclass in hype generation. By framing ChatGPT as an essential innovation, OpenAI has fueled public fascination with "AI," attracting millions of users and securing billions in investment. Beneath the spectacle, however, is, what appears to be, a calculated strategy led by Altman: capitalize on the "AI" boom, cement OpenAI's dominance, and ensure long-term profitability—far removed from the nonprofit ideals the company once espoused.

5—3—2

OpenAI's marketing of ChatGPT as a "general-purpose" "AI" capable of handling virtually any task creates the illusion of an all-powerful machine, while in reality, the technology remains heavily limited by its training data and algorithms. Altman's careful marketing strategy has sold the public a vision of an "AI-driven" future, where ChatGPT, with its natural-sounding responses and seemingly intelligent behavior, plays a central role. However, in practice, ChatGPT often falls short of its promises, offering responses that can be inaccurate, biased, or otherwise disconnected from reality. Despite these shortcomings, Altman and OpenAI continue to profit from the hype, capitalizing on the inflated expectations of a public that believes it's entered a new era of technology. Altman, as the puppet master behind OpenAI's strategic decisions, has leveraged the allure of "AI" to drive the company's rise into the ranks of Silicon Valley's most influential players. While his public image is one of a visionary leading the charge toward a utopian future powered by "AGI," his actions have consistently demonstrated that his true goal is financial success. By shifting OpenAI's focus from nonprofit ideals to corporate profiteering, Altman has orchestrated a lucrative transformation that relies on the perpetuation of "AI" hype and the cultivation of a public thirst for cutting-edge technology, all while reaping the rewards of a rapidly growing corporate empire.

Section 4 — Sundar Pichai: The Imitator General of Silicon Valley AI

5—4—1

In recent years, Sundar Pichai, CEO of Google and its parent company Alphabet, has spearheaded the company's urgent campaign to remain competitive with OpenAI, Microsoft, and other players in the so-called "AI" race. Central to Pichai's strategy has been the rebranding of nearly every Google product as "AI-powered." Features previously described as smart tools, automated enhancements, or algorithmic systems are now marketed as breakthroughs in "Artificial Intelligence"—even when the underlying technology remains largely unchanged. This semantic overhaul is not coincidental; it reflects Google's deliberate effort to align with market expectations and investor enthusiasm in a tech landscape overtaken by "AI" hype. Pichai's approach is strategic but reactive—aimed at defending market share rather than defining the frontier. This shift reached a pivotal moment with the post-ChatGPT launch of Google's Gemini platform. Following OpenAI's 2022 debut of

ChatGPT, Pichai issued an internal "code red," prompting resource reallocation and expedited product rollouts. The result was a series of hastily developed tools designed to signal Google's presence in the "AI" arms race. Yet, internal assessments soon revealed that Gemini lagged behind its competitors on multiple fronts, including adoption rates, performance, and user perception. Internal documents even acknowledged Gemini's failure to secure meaningful brand recognition—an embarrassing result for a company of Google's stature.

5—4—2

Google's rushed rollout of Gemini not only failed to impress but also generated controversy. The platform came under fire for its troubling image-generation behavior, which included the race-swapping of historical figures—an issue that critics argued undermined historical integrity and exposed latent ideological bias. The backlash highlighted a deeper problem: Google had prioritized speed over rigor, releasing a flagship product that was clearly not ready for wide deployment. While Pichai publicly acknowledged bias within the company's "AI" models, his response was vague and unsatisfying to many observers. Rather than offering substantive remediation or transparency, Pichai leaned on the idea that such technologies are "evolving" and require "ongoing refinement." For critics, this sounded less like responsible stewardship and more like deflection. The Gemini episode underscored Google's broader struggle: caught between its cautious corporate DNA and the market's demand for rapid innovation, the company has increasingly defaulted to conformity —mimicking the moves of its rivals rather than charting its own course. Perhaps most concerning, Gemini's behavior seemed to validate long-standing claims that Google embeds ideological assumptions into its supposedly neutral platforms. For years, critics have warned that Google exercises disproportionate influence over public discourse under the guise of technological neutrality. With Gemini, those concerns returned with new force: if the platform reflects the biases of the data and algorithms that trained it, and those systems originate from Google's ecosystem, then the implications extend beyond poor product design— they suggest an institutional posture, one that Pichai has chosen to neither refute nor reform.

Section 5 — Satya Nadella: The Mirage Investor

5—5—1

Under the leadership of Satya Nadella, Microsoft has firmly placed its bet on "AI," signaling a strategic pivot with its $10 billion investment in OpenAI in early 2023. This investment not only gave Microsoft a significant stake in the burgeoning field of "generative AI" but also aligned the company's vision with OpenAI's ambitious developments. The deal was framed as a partnership to advance the capabilities of "AI," allowing Microsoft to integrate OpenAI's models into its own product lineup—especially in the context of its search engine Bing and productivity software like Office 365. This investment came as part of a broader marketing push to rebrand Microsoft as an "AI-first" company. Nadella championed this vision in multiple public appearances, positioning "AI" as the core of Microsoft's future. The company's products, including Office applications like Word, Excel, and PowerPoint, were framed as being radically transformed by "AI" to enhance user productivity. Tools that had long been seen as a standard in productivity software were now being marketed as "AI-driven" solutions, promising smarter and more efficient workflows. However, despite such bold claims, much of the actual "AI" integration remains underdeveloped or lacks the breakthrough innovations Microsoft has previously indicated.

5—5—2

One of the key moments in this transformation was Microsoft's integration of OpenAI's ChatGPT into Bing, launching the product as a "better" search engine with the ability to provide generative, conversational responses. But this rush to adopt "AI" has not been not without its issues. Although Nadella and Microsoft have painted a picture of groundbreaking technology, there have been concerns over the limitations of these early "AI-first" offerings, particularly with regard to their practical applications and the fact that many of the promised functionalities are already available in other forms, such as through voice assistants like Siri or Google Assistant. Despite these critiques, Nadella's focus has remained firmly on positioning Microsoft at the forefront of "AI's" expansion. This focus is encapsulated in his ongoing rhetoric about how "AI" will change industries and transform workplaces. But as with many "AI" ventures, questions still remain about the real impact of these products. While the "AI" label has promised cutting-edge advancements, the underlying technology typically fails to live up to the hype, and its rush to deploy products without ensuring their maturity or reliability has left consumers and critics questioning the real value of such claims. At the heart of this strategy is the notion of "AI" being a catch-all solution—one that can be applied to virtually any aspect of Microsoft's offerings. Whether it is through smarter search engines, enhanced document creation tools, or improved "cloud" services, Microsoft seems to suggest that "AI" will usher in a new era of unprecedented innovation. However, much like the broader "AI-first" narrative, new promises come with the same old unfulfilled sense of wonder and hype, leaving consumers skeptical of the tangible benefits offered by these so-called "AI-powered" solutions.

Section 6 — Mark Zuckerberg: The Late Adopter Trying To Stay Relevant

5—6—1

Mark Zuckerberg, the co-founder and CEO of Meta Platforms, Inc. (formerly Facebook, Inc.), has long been a controversial figure in the tech world. The influence he has had over social media and the tech industry is undeniable, but so too is the scrutiny he has faced over issues such as privacy concerns, data misuse, and Meta's handling of misinformation. More recently, Zuckerberg has attempted to stay relevant in the rapidly changing tech landscape by shifting from his failed Metaverse vision to promoting the so-called "AI revolution."

5—6—2

In late 2021, Meta made a bold and highly publicized shift under Zuckerberg's leadership: the company rebranded from Facebook, Inc. to Meta Platforms, Inc.—more commonly known as just Meta—and declared its dedication to building the so-called "Metaverse." This was a costly and ambitious venture designed to position Meta as the leader in the next wave of virtual reality and augmented reality technologies. Zuckerberg, in particular, championed the Metaverse as the future of the Internet, promising fully immersive virtual spaces where people could work, socialize, and interact in unprecedented ways. However, the project failed to materialize as promised. The vast sums of money poured into virtual and augmented reality technologies did not translate into widespread adoption. Meta's investments in VR hardware, such as the Oculus headsets, struggled to gain significant traction, and the Metaverse itself remained a distant and underwhelming prospect. Despite this, Zuckerberg continued to double down on his vision, insisting that the Metaverse would eventually reshape the digital world. However, the reality was starkly different. As investor skepticism grew, Meta's stock took a significant hit. By 2023, the company had pivoted away from the Metaverse and swiftly repositioned itself as a leader in the so-called "AI revolution." With the rise of "AI" hype —particularly after OpenAI's ChatGPT gained widespread attention—Zuckerberg quickly reframed Meta's priorities. The company began integrating "AI" into its social media platforms, using it for content recommendations, ad targeting, and user engagement. As a result, Meta's marketing strategy in 2023 leaned heavily into "AI" as its next defining feature. However, as with the Metaverse, concerns quickly emerged. While

Zuckerberg presented Meta as being at the forefront of "AI" innovation, critics argued that the company's sudden embrace of "AI" was merely a reaction to the latest industry trend, rather than a sign of genuine technological leadership.

5—6—3

Beyond business strategy pivots, Zuckerberg's leadership has continued to stir controversy. Though it may not come as a surprise to many, Meta is still notorious for its repeated mishandling of user data, which includes the infamous Cambridge Analytica scandal. This was the one wherein it was revealed that the personal data of millions of Facebook users had been improperly accessed and used for political targeting. Zuckerberg's public image was severely damaged by it, and Meta has since faced multiple regulatory investigations related to data privacy practices in both the United States and the European Union. Additionally, Zuckerberg's management style has often been described as aggressive and top-down, with critics accusing him of prioritizing profit over user well-being. Rapid changes to the platform—such as prioritizing short-form video and "AI-driven" content curation—have alienated longtime users accustomed to Facebook's traditional format. Meanwhile, Meta's content moderation remains a point of contention, as the company faces accusations of both suppressing certain political viewpoints and allowing harmful content to spread unchecked. Despite the controversies, Zuckerberg remains determined to position Meta as a leader in the "AI" space. Much like the earlier Metaverse pivot, this new strategy seems less about innovation and more about ensuring Meta's continued dominance. However, Meta's ongoing struggles with data privacy, misinformation, and regulatory scrutiny suggest that "AI" will not resolve the company's deeper systemic issues.

5—6—4

Zuckerberg's abrupt transition from the Metaverse to "AI" reflects a pattern of reactive decision-making, driven by the need to stay relevant in an evolving tech landscape. His shift from virtual and augmented reality to "Artificial Intelligence" was not part of a carefully planned long-term vision, but rather a response to prevailing industry hype. Just as he once positioned the Metaverse as the future, he now claims that "AI" is Meta's next great transformation. However, much like the Metaverse, his focus on "AI" lacks tangible, transformative outcomes. This attempt to rebrand Meta as an "AI-first" company appears to be a superficial fix rather than a strategic reinvention. Meanwhile, the company continues to grapple with unresolved issues—data privacy violations, misinformation, and public distrust—that "AI" will not magically solve. Despite Zuckerberg's persistent efforts to recast himself and Meta as cutting-edge technology pioneers, his track record of chasing industry trends without delivering lasting value leaves many skeptical. Given the increasing public scrutiny of the exaggerated and often misleading narratives surrounding so-called "Artificial Intelligence" across the tech industry, it seems likely that Meta's latest pivot will ultimately fail to produce the revolutionary impact Zuckerberg claims it will.

Section 7 — Other Key Figures Perpetuating the Fraud

5—7—1

While figures like Elon Musk, Sam Altman, and Satya Nadella have been prominent players in the rush to capitalize on the fraudulent "AI" narrative, there are several other key figures and entities in the tech world who have contributed to the amplification of the fraudulent narrative. From the GPU dominance of NVIDIA to the overblown promises of human-level intelligence from DeepMind, many continue to push the idea of "AI" as a "transformative technology" despite its inherent fundamental limitations. This also includes the tech media, which regularly hypes the narrative without proper scrutiny, further fueling the illusion of imminent breakthroughs.

5—7—2

Jensen Huang—As the CEO of NVIDIA, Huang has played a pivotal role in amplifying the "AI" hype, leveraging the company's dominant position in the GPU market to promote the narrative of "AI" as a sweeping technological revolution. NVIDIA's graphics processing units (GPUs) have indeed become the hardware of choice for training the massive, resource-intensive models that underpin what is being marketed as "machine learning." Huang has consistently positioned NVIDIA as essential to the future of so-called "AI" technology, promoting the view that GPUs are central to breakthroughs in everything from healthcare to "autonomous" driving. While GPUs are unquestionably powerful tools for scaling the computational demands of today's "machine learning" systems, the broader claim that these systems constitute true "Artificial Intelligence" remains highly debatable. Huang's public messaging frequently blurs the distinction between the hardware driving statistical pattern recognition and the notion of actual machine intelligence. By framing GPU demand as synonymous with progress in "AI," Huang has benefited from a market saturated with investment in what many critics view as an inflated and largely unfulfilled sector. This positioning has allowed NVIDIA to flourish as enthusiasm for "AI" continues to rise, despite ongoing concerns about the limited scope and capabilities of these systems. Huang's frequent predictions that "AI" will soon be a seamless part of everyday life have been met with growing skepticism. Critics argue that these claims overstate the technology's current capabilities and ignore the significant conceptual and technical gaps that separate today's models from anything resembling general intelligence or human-like understanding. While NVIDIA's GPUs are unquestionably integral to the current infrastructure of what is branded as "machine learning," the leap from scalable computation to actual intelligence remains unresolved. In this context, Huang's framing of NVIDIA's hardware as the backbone of the "AI" future reflects a broader trend of misrepresenting the nature and potential of these technologies.

5—7—3

Demis Hassabis—As co-founder and CEO of DeepMind, Hassabis is yet another central figure in the "AI" hype machine. Since DeepMind's early claims of breakthroughs in reinforcement learning and its media-exaggerated "victory" in the game of Go, Hassabis has positioned the company as a supposed leader in the quest for "Artificial General Intelligence" (AGI)—a form of "AI" that is said to replicate human-like cognition across a broad range of tasks. However, DeepMind's public communications have repeatedly overstated the timeline for achieving "AGI," frequently suggesting the company was nearing systems capable of human-level intelligence. Despite DeepMind's hyped results in highly specialized tasks—like AlphaGo and AlphaFold, which were framed as "AI achievements"—these claims remain unsupported by any evidence that demonstrates real-world generalization beyond narrow-task design. Such tasks were narrowly defined challenges, with no applicability beyond specific, constrained use cases. These repeated suggestions of imminent human-level intelligence are not just premature—they present a deeply misleading view of current capabilities. DeepMind's ongoing branding around "AGI," coupled with its efforts to repackage narrow systems as groundbreaking, plays a central role in perpetuating the broader, fraudulent industry narrative about "AI" capabilities. The gap between these narrow, specialized systems and actual "AGI" is so vast that any claim of near-term development is baseless. The dream of "AGI"—like much of the so-called "AI" hype—remains a fabricated illusion perpetuated by industry marketing rather than technological substance.

5—7—4

The Tech Media—And while Huang and Hassabis have certainly earned their "honorable mentions" as key players in the perpetuation of the fraudulent "AI" narrative, I would be remiss not to acknowledge the tech media's crucial role in amplifying these claims without the necessary skepticism or scrutiny. Numerous tech outlets have routinely embraced the idea that "AI" is currently transforming society, often reporting on so-called advancements in "machine learning" as though they represent significant and credible steps toward "Artificial General Intelligence" (AGI). Articles frequently blur the line between sophisticated computational systems and actual intelligence, encouraging the public to believe that genuine "AI" has already arrived. This widespread narrative inflation, driven by the tech media, has distorted public expectations and helped companies profit from the illusion that they are leading an "AI revolution," regardless of the actual substance behind their products or services. Headlines touting "AI-powered" assistants, "autonomous" systems, and human-like chatbots reinforce the belief that we've entered a new technological epoch. In truth, the majority of these so-called advancements are marginal improvements in statistical modeling and pattern recognition— nowhere near the fantastical outcomes implied by media coverage. The tech media's enabling role in this narrative cannot be overlooked. By serving more as promotional conduits than investigative watchdogs, many journalists and outlets have enabled figures like Musk, Altman, Huang, and Hassabis to promote their self-serving visions with little resistance. Unless there is a meaningful course correction in how these narratives are reported, the deceptive "AI" hype cycle will continue misleading consumers, investors, and policymakers alike.

"AI" Means Fraud

"Falsehood flies, and the Truth comes limping after it."

Jonathan Swift
The Examiner No. XIV, 1710

"I compare it with a lie, which like to a snowball, the longer it is rolled the greater it becomes."

Martin Luther
The Table Talk of Martin Luther, 1967

Chapter 6:
The Consequences of the Lie
(The Real-World Impact of the
Hype)

Section 1 — The Economic Impact: Billions in Investments and False Promises

6—1—1

The widespread narrative surrounding so-called "AI" has driven unprecedented economic, social, and legislative shifts. The short-term economic benefits—from soaring investments to the rapid expansion of data infrastructure—have fueled excitement and reinforced the illusion that this technology is revolutionizing every industry it touches. However, as we've already considered, this so-called technological boom is built on overblown and, in many cases, outright deceptive promises—promises that will likely lead to long-term consequences that are far less favorable. In this chapter, we'll begin by examining the economic impacts—both short-term and long-term—of the fraudulent "AI" boom. The initial financial windfall has led to a surge in capital investments, corporate expansion, and a reshaping of educational and labor trends. Yet, beneath this temporary prosperity lies a fragile foundation built more on perception and marketing than on demonstrable capability. When public perception inevitably shifts—when the nature of the technology and its true limitations become undeniable—the long-term fallout could be immense, echoing past technological disillusionments that left industries upended and countless people in financial ruin.

The short-term financial impact of the "AI" narrative has, for now, been overwhelmingly positive. Billions of dollars have flooded into companies claiming to be at the forefront of this so-called "AI revolution," and stock prices for companies even loosely associated with "AI" have surged. NVIDIA, for example, experienced record growth as demand skyrocketed for its GPUs, which are aggressively marketed as essential for training "AI" models. The company briefly crossed a $1 trillion valuation in 2023, driven not by a revolution in capability, but by the illusion that "AI" is redefining modern life. What's actually being redefined is the term "Artificial Intelligence" itself. At the same time, universities have seen a spike in Computer Science enrollments, fueled by the promise of lucrative careers in "AI" and machine learning. According to the Computing Research Association's Taulbee Survey, student numbers have climbed notably, with institutions scrambling to expand course offerings to meet demand. Corporations, eager to maintain the illusion of explosive progress, have hired aggressively, and corporate spending on "AI" initiatives has reached staggering levels. A McKinsey report estimated that over $300 billion was spent globally on "AI" in 2023, including investments in so-called "AI-powered" customer service, predictive analytics, and automation platforms. What's unfolding is a self-reinforcing cycle: the more capital that flows in, the more convincing the illusion becomes. However, like any bubble built on hype, the question is not *if* reality will catch up—it's *when*.

Despite the gold rush mentality of the moment, the long-term economic outlook tells a very different story. We've seen this play out before: unsustainable hype propped up by unrealistic claims, leading to widespread disillusionment once the promises are exposed as hollow. A central issue is the mounting infrastructure cost required to keep up the illusion of progress. Massive data centers are being built to feed the computational demands of large "AI" models—facilities that consume enormous quantities of energy, water, and hardware. According to the International Energy Agency, data centers already account for nearly 1% of global electricity consumption, and that number is poised to rise dramatically with increased "AI" workloads, alongside cryptocurrencies and the broader digitization of society. These costs are spiraling upward, even as the tangible returns— economic, functional, or societal—remain questionable. This model is unsustainable. Once the limitations of "AI" become undeniable to the general public and investors alike, sentiment will shift. We've seen the pattern before, most notably during the dot-com burst: inflated valuations based on speculation, media hype, and smoke-and-mirror promises—followed by implosion. The difference this time is that the scale is larger, the infrastructure more expensive, and the public narrative more deeply entrenched. Universities that rushed to cash in on the "AI" boom may soon find themselves rethinking their entire academic strategy. Students lured by the promise of stability and relevance may find a job market far more limited than advertised. And policymakers—some of whom have drafted legislation around the assumption that "AI" is inevitable—may be forced into backpedaling, revising policies built on hype rather than substance. And let's be clear about something: this isn't just about market cycles or benign optimism. It's about deliberate distortion. One cannot call people like Elon Musk or Sam Altman "technology experts" while simultaneously claiming they couldn't possibly have known better. Either they understand the technologies they're promoting—and have knowingly exaggerated the capabilities of such technologies—or they don't, in which case they were never qualified to be called experts in the first place. You can't have it both ways.

6—1—4

In summary, while the short-term economic boom has been substantial, it rests on a shaky foundation—one constructed from marketing sleight-of-hand, technical misrepresentation, and strategic exaggeration. If history is any guide, the long-term consequences could be far more damaging than most are prepared to admit. The problem with technological bubbles isn't merely that they burst—it's that, when they do, they expose institutional failures: regulatory complacency, corporate irresponsibility, and a lack of meaningful public scrutiny. The fraud surrounding today's "AI" narrative is not just a misjudgment—it's a calculated exploitation of public ignorance, investor greed, and legislative inertia. As such, when the fallout does arrive, it will not only hit balance sheets—it will hit careers, institutions, and public trust. Recognizing this now —rather than after the consequences unfold—is the only way to mitigate the worst of what is possibly yet to come.

Section 2 — The Mental Health Impact: How the Fear of "AI" and Its Imagined Capabilities Are Negatively Affecting Mental Well-being

6—2—1

Beyond the economic impacts, the fraudulent "AI" narrative reaches further—into the minds of the public, distorting perception and destabilizing mental well-being. Yes, mental health is shaped by a multitude of factors, and no sane society can—or should—attempt to bubble-wrap every psyche.

However, this is different. This isn't the result of ideological tension or legitimate scientific debate. The fear and anxiety surrounding "AI" stem not from uncertainty about a real threat, but from a calculated campaign of deception—engineered for financial and political gain. This isn't a case of public overreaction to an emerging technology. It's fear as a consequence of deliberate exaggeration, distortion, and relentless hype. From phantom threats of omnipotent machines to farcical predictions of jobless dystopias, the public is being bludgeoned with misinformation. As a result, people are breaking under it—psychologically, emotionally, and existentially. I'm not a psychologist and won't pretend to offer diagnoses, but I am offering reasoned observations rooted in patterns that are now impossible to ignore: the fraudulent "AI" narrative is not only corrupting our economic systems and institutions—it is corroding mental health on a notable scale.

6—2—2

One of the most pervasive fears infecting public consciousness is that "AI" will wipe out jobs on a mass scale. The narrative suggests that entire sectors—manufacturing, retail, logistics, customer service, even software development—will be gutted, leaving millions unemployed and directionless. Yet, this isn't a logical forecast. It's a scripted panic, echoed endlessly by media cycles that trade nuance for clickbait. For one thing, automation is nothing new. For another, jobs have always evolved with every industrial shift in history. Lamplighters vanished with electricity. Switchboard operators disappeared with digital telecom. In each case, new industries, skills, and opportunities emerged. However, the fear around "AI" isn't a natural reaction to progress—it's a symptom of manipulated perception. The difference this time is that the panic isn't trailing real innovation—it's preceding it, built on fantasy and sold as fact. Yes, automation does reshape labor markets, but the apocalyptic tone surrounding "AI" isn't a byproduct of genuine technological disruption—it's a marketing campaign. And while the architects of this narrative may not have directly intended to incite mass anxiety, that's exactly what they've done. The resulting mental toll is a predictable consequence of fraud.

6—2—3

Beyond fears of losing their livelihoods, many people are haunted by something more abstract: the idea that "AI" will become sentient, autonomous, and ultimately uncontrollable. This fear, born of science fiction and nurtured by Silicon Valley marketing, envisions machines rising to replace not only our labor, but our authority—our ability to choose, judge, create, and govern. It is a fear rooted in a false dichotomy: human control versus machine control. The truth is simpler and uglier—there is no machine control. There is only human manipulation by those wielding the tools. Let's get this straight: the systems being touted as "AI" today are not alive. They are not aware. They are statistical models operating within hard-coded parameters. They do not think. They do not understand. They do not want anything. And yet, public figures, media outlets, researchers, and corporations routinely conflate predictive algorithms with cognition—feeding public dread while elevating their own status. The existential fear of sentient machines is not the result of an actual trajectory of innovation. It is a psychological side-effect of persistent misrepresentation, weaponized for profit and control. The real danger here isn't that machines will develop minds of their own—it's that those in power will use these tools to magnify their own reach, eliminate accountability, and offload decisions that ought to require moral judgment onto systems designed to optimize for speed and efficiency. The people wielding these tools are using a false narrative to justify moral abdication. They're insisting, "The technology could do it," when in truth, anything the technology does is because people built it to do exactly what it did. "AI" doesn't eliminate accountability—it diffuses it. And that's what makes it dangerous. The machines aren't replacing our agency—we are surrendering it.

6—2—4

As if the fear of sentient machines weren't unhinged enough, a small but vocal contingency has begun pushing for legal rights for "AI." You read that correctly. There are people arguing that these systems—trained on scraped data, operating with zero consciousness or intent—should be granted rights traditionally reserved for living beings. They propose legal personhood for pattern-matching algorithms. Let's pause and think about just how absurd this is. Suppose for a moment that "AI" is not only a legitimate technological revolution but is also truly sentient and autonomous. Now imagine receiving a court notice informing you that your mobile phone is suing you for "unlawful detention" because you've kept it in your pocket too long. Meanwhile, after yelling at Siri for giving you the wrong weather forecast, it indignantly insists that every such outburst constitutes wrongful accusation without due process. If you find that all laughably absurd, it's because that's precisely what it is. This isn't a harmless eccentricity. It's fundamentally erroneous —an engineered distortion of what rights are and who or what can possess them. These aren't sentient beings. They're programs trained on data sets. They have no inner lives. No consciousness.

No pain, no joy, no capacity for experience whatsoever. Granting rights to such tools is not progressive. It's regressive. It's an attack on the meaning of rights themselves. Yet, this absurd idea is being legitimized by tech insiders, ethicists, and academics who either sincerely believe these systems are approaching consciousness—or find it convenient to act like they do. In either case, it's a scam: a philosophical circus act meant to shift attention away from real, present-day harms. While their are people suffering from fear of unemployment or loss of control, and the erosion of institutional accountability, the same people selling the "AI revolution" are trying to convince us that computer systems deserve our empathy. Why? It's because framing software as morally significant entities sets the stage for legal maneuvering, tax avoidance, liability shielding, and unearned legitimacy. It turns "AI" from a product into a stakeholder. From a tool into a partner. From a liability into a victim. And as cold as it may be of me to say so, if we let this happen then we deserve every bit of the institutional breakdown and degradation of humanity that follow.

Section 3 — The Impact on Legislation: How Laws Are Being Created Based on False Premises

6—3—1

Another prominent consequence of the fraudulent narrative behind so-called "AI" technology has manifested itself as a growing trend in U.S. legislation—at both the state and federal levels—to regulate "AI" as though it is a unique, legitimate technology requiring unique, new legislation. This stems from a widespread misconception that "AI" is somehow fundamentally different from existing automation technologies. However, as I hammered down in Chapter 3, what's often labeled as "AI" is really just some combination of automation and other existing technologies—albeit sometimes souped-up with more data and processing power—that have been in use for decades. The reality is that while there are advancements in scale and capability, there is no fundamentally new type of technology here— nothing that inherently requires an entirely separate legislative category. Many existing legal frameworks, such as anti-discrimination laws and consumer protection statutes, already inherently address a majority of the risks these automated systems present. Moreover, it is vital to note that all such enforcement efforts ought to focus on holding people—sentient, autonomous beings—

accountable for their misuse or fraudulent misrepresentation of the technology, not the technology itself, which possesses no sentience or autonomy. As I wasn't claiming to be a psychology professional in the previous section, I am also not claiming to be a legal professional in this section. What follows, again, are my own reasoned observations.

6—3—2

Laws like the *Colorado AI Act* (2024), which target "high-risk AI systems" on matters like discrimination, are based on the false premise that "AI" is an actual, fundamentally different kind of technology. In truth, these automated systems are simply more sophisticated versions of the same automation tools that have been regulated by existing laws for years. For example, discrimination laws, like the Civil Rights Act of 1964, already apply to the use of automated systems that may inadvertently cause illegal bias. The same notion applies to deceptive practices, such as in marketing, which are already covered by FTC regulations. These existing laws are inherently equipped to address most of the risks posed by the progressing use of automation. In many cases, the challenge lies more in enforcement than in legislative gaps— meaning new "AI"-specific laws are often redundant rather than necessary. The false distinction between "AI" and other technologies is dangerous because it leads to unnecessary regulation, which, in turn, creates confusion and unnecessary complexity. Legislators are being misled into thinking that they need new laws to govern something that doesn't actually exist. In reality, the risks we are concerned with—discrimination, privacy violations, and fraud—are not new or specific to "AI," and they are already regulated under existing laws.

6—3—3

You might be wondering where this trend of believing that new "AI"-focused legislation is required originates from. Well, as I see it, as with all of the other consequences of the fraudulent "AI" narrative, it is fueled by fear and misinformation rather than a clear understanding of the actual technology itself. The same public bombarded with deceptive claims about the capabilities of "AI"— such as the idea that it is, or could become, sentient and take control, or that it might make autonomous decisions with catastrophic consequences— includes the very same lawmakers pushing for these redundant laws. These fears are based on fiction and are stoked by the media and high-profile public figures. Take, for example, the reaction to "AI" in the form of nuclear ban bills (2023-2024). These bills are driven by the fear that "AI" could somehow gain control of weapons systems or other critical infrastructure, but that fear is based on the false belief that "AI" is a lot more than it actually is. As discussed earlier, though, "AI" is nothing more than a modern presentation of automation, algorithms, and other existing technologies working within strict human-defined parameters—there is no sentience nor independent decision-making. If this technology could gain control of weapons systems or other critical infrastructure, it would be because the humans behind it programmed and leveraged it to do so. Then again, it would also be because the humans responsible for the oversight and maintenance of such weapons systems or other critical infrastructure failed to detect and stop such an operation before the takeover was completed. It wouldn't be because so-called "AI" technology made independent decisions or took independent actions to make such a takeover a reality. So, with that in mind, it bears repeating that the fears related to something like that ever happening are rooted in the hype and misinformation embedded in the Hollywood-style narrative behind "AI," not the reality of the technology's capabilities. Even laws that seem more practical, such as the 2023 update of the California Privacy Protection Act (CPPA), which added language around "automated decision-making," reflect the assumption that such systems pose new and distinct threats. In reality, most of the risks being addressed—like privacy violations—were already covered under the 2018 version, and the changes are arguably more reflective of scale and public pressure than of genuinely new technological threats. The real issue is not automation itself, but how companies or people use it—and those practices are already largely regulated. The push for new "AI"-specific laws reflects an exaggerated response to problems that have been around for years. In short, many "AI" laws are a direct response to the inflated narrative surrounding these technologies, fueled by fear and the desire for dramatic action. Policymakers, influenced by public fear and media hype, are crafting laws that target something that doesn't actually exist as it is portrayed. They are chasing a problem that doesn't need a new solution, all because they've been misled into thinking "AI" is something more than what it really is: just arguably souped-up automation. Then again, I wouldn't dismiss the notion that some of them have self-serving ulterior motives behind their calls for or support of such legislation.

Section 4 — Public Perception and Trust in Technology: How Misleading Narratives About "AI" Damage Public Trust in Technology

6—4—1

The final section of this chapter on the negative consequences of the fraudulent narrative behind so-called "AI" technology considers its impact on public perception and trust in technology. The hype and misinformation surrounding the capabilities of "AI" have led to widespread skepticism and wariness among the public. As the tech industry continues to push fraudulent claims about "AI" as an all-powerful, "autonomous" force—whether through marketing, media manipulation, or outright deception—the consequences extend far beyond this particular subset of technology. The damage is systemic, chipping away at public confidence in technological advancements in general. As a result, public trust in technology as a whole suffers. A key issue is that many people who interact with so-called "Artificial Intelligence" technologies are often unaware of the true nature of the technology they're engaging with. As a result, they may worry about potential job loss, privacy violations, or even the possibility—falsely marketed though it may be—of technological systems making life-altering decisions without human oversight.

As a further consequence, when people are fearful of technology, they are less likely to embrace innovations that could improve their lives, reduce their workload, and even enhance their well-being. Instead, they may resist adopting new technologies, either because they don't understand them or because they perceive them as too risky. The result is a drag on technological progress (ironic, right?)—both for businesses that rely on such technologies and for individuals who could benefit from them. This very scenario is playing out with "AI" technology. Repeated exposure to unrealistic portrayals of "AI" technology—depicting it as self-aware, independently capable, or poised to replace or even consistently compete with human intelligence—has fostered a culture of distrust. This is not accidental. As I have indicated repeatedly, it is the direct result of an industry more invested in securing funding, market dominance, and regulatory advantages than in providing a transparent, accurate understanding of the technologies it markets. The public is caught in a cycle of overhyped promises, inevitable disappointments, and reactionary skepticism. People have been led to believe that the technologies being fraudulently marketed as "Artificial Intelligence" are either an omnipotent threat or a utopian solution—both of which obscure the reality: that they are neither.

Compounding this problem is the role of media and pop culture, which have long thrived on sensationalized narratives, particularly when fed by industry PR. The constant bombardment of dystopian warnings, exaggerated success stories, and misinformed commentary ensures that the public is fed a steady diet of misinformation. Few take the time to dissect what is actually being sold to them, and fewer still recognize the vested interests behind these narratives. Instead of fostering an informed discussion about technology's real capabilities and limitations, the conversation is driven by fear-mongering, marketing spin, and the self-serving agendas of those who stand to profit from the confusion. Moreover, this misunderstanding and fear of "AI" also opens the door to more extreme, populist, and reactionary responses. Policymakers, driven by public fear, may impose unnecessary regulations or, worse, create misguided laws that hinder legitimate technological advancement without addressing the actual concerns at hand. This only further perpetuates the cycle of distrust, as consumers and businesses alike struggle to navigate a regulatory environment that is largely reactive rather than thoughtful or balanced.

At its core, the public's skepticism towards the technology industry is not just about "AI"—it is about the integrity of technological discourse. If even consistent, rudimentary trust in technology is to be established, the cycle of deception must be broken. This means cutting through the noise, exposing the fraudulent narratives for what they are, and demanding transparency from those who seek to control the conversation. And yes, it's important to emphasize that public trust in technology is built on transparency, education, and a clear understanding of the capabilities and limitations of these tools. When technology companies present their products with honesty and clarity, focusing on real-world applications and practical benefits rather than inflated promises, the public is more likely to embrace and trust those innovations. By properly framing "AI" as an arguably modernized set of existing tools that are ultimately governed by human control and oversight, society can begin to see it for what it truly is—a powerful, though flawed, tool, not a mysterious or omnipotent force. There is also a need for better media literacy, as the sensationalized narratives about "AI" that dominate headlines must be countered with factual, balanced information. Technology experts have a responsibility to educate the public about what technology can and cannot do. By highlighting existing regulations, standards, and ethical guidelines that govern the development and deployment of technologies, we can create a sense of control and accountability to the conversation. Until then, public skepticism is not just justified—it is essential for resisting the industry's systemic dishonesty.

"The truth is more important than the facts."

Frank Lloyd Wright
The Autobiography of Frank Lloyd Wright, 1932

"The truth will set you free."

Jesus Christ
The Holy Bible, John 8:32

Conclusion

Section 1 — Where This Is Headed: The Future of the "AI" Hype Cycle

C—1—1

As I detailed in Chapter 1, Section 3, history provides clear examples of the "AI" sector following a predictable pattern—one of exaggerated claims, misplaced enthusiasm, and inevitable collapse. These periods, often referred to as "AI Winters," have occurred multiple times, each one driven by a cycle of overpromising, underdelivering, and ultimately disappointing those who bought into the hype. It's worth reiterating this here because this recurring pattern serves as the perfect foundation for understanding where the present "AI" bubble—if you will—is headed. The first major "AI Winter" in the 1970s followed the realization that "symbolic AI"—early attempts to simulate human reasoning—was fundamentally flawed and incapable of achieving its overly ambitious goals. The second, in the late 1980s and early 1990s, came after expert systems, once hailed as the future of intelligent computing, collapsed under their own inherent flaws and impracticality. A third, partial "AI Winter," though less widely recognized, emerged in the early 2000s as disillusionment set in with "machine learning" and "neural networks," which—despite their mathematical complexity—failed to deliver on their promises of broad, real-world intelligence. Now, we find ourselves in yet another cycle of "AI" hype

—except this time, the deception is even grander in scale. Unlike previous iterations, which were largely driven by researchers making overzealous claims, today's "AI" hysteria is a marketing-driven spectacle. Corporations, media outlets, and investors have all bought into the illusion, propping up an industry that thrives on misleading terminology and exaggerated capabilities. But as history has shown, such deception is not sustainable. The question is not whether this cycle will collapse, but *how* it will collapse.

C—1—2

From here, I see three primary ways in which the present "AI" hype bubble could burst (or the cycle could end):

- **Scenario 1: The Traditional "AI Winter" (History Repeats Itself)** — In this scenario, the collapse unfolds much like it has in the past. The industry continues pushing its fraudulent claims, but as the limitations of the technology become more apparent, frustration grows. Companies fail to deliver on their promises, investors begin to lose faith, and the money that once fueled this "AI" boom starts to dry up. Governments, once eager to regulate and invest in "AI," shift their focus elsewhere as they realize they've been sold a fantasy. Eventually, public enthusiasm wanes, and the next "AI Winter" sets in. This would be the most historically consistent outcome.

- **Scenario 2: The Grift Persists Indefinitely (The Deception Continues)** — Alternatively, the current cycle may not collapse in the traditional sense—at least not for a long time. The tech industry has become more sophisticated in its ability to sustain hype, thanks to media complicity, regulatory lobbying, and relentless rebranding. If the fraudulent narrative surrounding "AI" remains profitable and useful to those in power, they may find ways to keep the illusion alive indefinitely. By constantly moving the goalposts—shifting from "AI will replace all jobs" to "AI will enhance jobs", for example—the industry could maintain enough momentum to prevent outright collapse. This would result in a slow, drawn-out disillusionment, if even that, but one that never fully materializes into a proper reckoning.

- **Scenario 3: The Voice of Reason Breaks Through (The Curtain is Torn)** — In this scenario, enough people wake up to the fraud before another "AI Winter" is needed to force the truth into the open. While this would be the most direct path to dismantling the deception on a more longterm basis, it would also pose the greatest threat to those who have built their influence, business models, and reputations around the fraudulent "AI" narrative—not to mention the countless students around the world who've invested time and money in pursuing degrees tailored to "AI" technology. Unlike past "AI" collapses, which were driven primarily by present technological shortcomings, this scenario would represent a complete ideological shift—one that could cause irreparable damage to the perception of "AI" as a legitimate technological frontier. Still, for those who value truth and accuracy, such as myself, this would be the ideal outcome, as it would finally tear the curtain (referring back to my earlier *Wizard of Oz* parallel) and expose the truth of so-called "Artificial Intelligence" once and for all.

141

While history suggests that **Scenario 1** is the most likely to happen, the increasing sophistication of the tech industry's propaganda machine means **Scenario 2** cannot be ruled out entirely. However, if enough people recognize what's really happening, **Scenario 3** remains a genuine possibility—especially if more experts step forward and tirelessly expose the truth. To that end, the question now is whether the public will allow itself to be fooled yet again—or whether, this time, the truth will prevail before the cycle can repeat itself.

Section 2 — How to See Through the Lies: Recognizing Marketing Tricks and Deceptive Language

C—2—1

The tech industry—especially the so-called "AI" sector—thrives on deception. It's full of slick marketing tricks designed to fool the average consumer. If you find yourself frustrated or even feeling betrayed by the technology industry, especially all of the constant "AI" hype, you're not alone. In this section, I'm going to arm you with the knowledge you need to spot the most common tricks: vague language, unnecessary (often meaningless) buzzwords, overblown promises, secretive black boxes, fake or compromised experts, and the constant refrain that the future is already here. Once you understand these tricks, you'll never observe technology hype the same way again. Let's dive in, starting with one of the sneakiest tricks: vague and ambiguous terminology.

C—2—2

Vague and Ambiguous Terminology—The tech industry loves using vague, meaningless terms that sound impressive but are essentially smoke and mirrors. Take "Artificial Intelligence"—it's been marketed to death, redefined so many times that it shouldn't mean anything anymore. Then, there's "the cloud." It's been sold as some magical, futuristic concept, but let's be real here—who actually thinks a fluffy cloud stores your photos? "The cloud" is just somebody else's computers doing work for you in a shared-effort manner, a technology we've had for decades. The tech world just slapped a shiny new name on it to get people excited in the consumer market. And "AI" works the same way, because instead of saying, "This is a complex computer program that matches patterns," companies say, "Our AI understands you!" It's all about making the technology sound more impressive than it really is.

Here's how to avoid falling for this trick: When you hear terms like "AI" or "the cloud," ask for a simple explanation of what it really means. If they can't give you one—or if the explanation sounds awfully fanciful or hard to believe—it's probably just hype. When vague, ambitious terms aren't enough to sell the narrative, buzzwords come to the rescue.

C—2—3

Buzzwords—These vague terms are bad enough, but the industry takes it even further with buzzwords that sound technical but are often unnecessary and/or meaningless. Buzzwords are like the cousins of vague language—they sound technical and impressive, but they're just as empty. Terms like "neural networks," "deep learning," and "big data" get thrown around to make tech seem revolutionary, even when it's not. Granted, these words aren't always misleading by themselves, but they're often used to confuse and dazzle you, rather than actually inform you. The average person doesn't have the tech background to see through the jargon, so they're left thinking these are groundbreaking ideas. In reality, they're often just rebranded versions of old tech with a few tweaks. For instance, "neural networks" and "deep learning" are marketed as if they mimic human thinking, but they don't. They're just algorithms matching patterns in data. In one sense, it's like a librarian guessing your next book based on what you've borrowed before. It's not "learning" like a human brain—it's just finding patterns. It's not the "intelligence" it's claimed to be. The goal isn't to explain the tech—it's to overwhelm you with complexity so that, in your state of awe, you don't ask the right questions, which could shatter your sense of awe and wonder.

Don't fall for it: If you hear a buzzword, like "deep learning," ask for a simple explanation of what the term actually means and contributes to the technology it's attached to. If the explanation equates out to something much less impressive than the word itself, such as matching patterns, as "deep learning" is, then what you've come across is likely a buzzword. Buzzwords create confusion, but another trick is even worse—overpromising what the technology can actually do.

C—2—4

Overpromising and Under-Delivering—One of the oldest tricks in marketing anything is making bold, unrealistic claims about what a product can do—claims that sound too good to be true, because they usually are. Tech companies love to promise the moon: "AI will revolutionize every industry" or "Our AI can reliably and consistently predict human behavior." However, when the product actually arrives, it's a massive letdown—failing to live up to the hype. Take "self-driving" cars, for example. Companies like Tesla promised "full autonomy by 2024," but we're still seeing excessive crashes and human drivers needed. It's the same trick: overpromise to stir excitement and get funding, then under-deliver, leaving you disappointed. It's this very kind of hype that fuels the fears we talked about in Chapter 6—people start worrying about job loss or privacy violations that never come to pass, making the public even more distrustful.

Here's how to spot it: Check the track record of the person or company making a claim. Do they have an established practice of making big claims, but then not delivering on them? Do they tend to make a lot of excuses when the actual technology fails to match the promises of its marketing? If so, don't buy into the hype. In any case, proceed with great caution, especially if all of the marketing presents the product as if it were some sort of mysterious black box.

C—2—5

The Black Box Fallacy—Overpromising is one thing, but the next trick we'll consider keeps you from asking any questions at all—the "black box" fallacy. Some companies love to hide what their technology actually does by referring to it as if it were a mysterious "black box"—too complex for you to understand. They do this all the time with "AI," making it sound like some mysterious, magical system that's beyond the comprehension of the average John and Jane Doe. In many cases, there's a shred of truth to this—though less so for those who've been reading this book. In reality, though, most technology boils down to fancy math —complex in some cases, but not impossible to understand on a fundamental level, and usually nothing all that groundbreaking. Again, consider Tesla's "self-driving" technology. They market it as "advanced AI technology," but they never seem to explain how it works, hoping you'll trust its marketing blindly, even after some 736 crashes reported by NHTSA in 2024. It's like a car salesman saying, "This engine's too high-tech to explain—just trust me, it's amazing!" Then, the car breaks down a month or so later leaving you with a repair bill that's much higher than you could have ever imagined. They're typically hiding something, and you shouldn't just trust them. By making the technology seem mysterious and overly-complex, companies hope to stop you from asking reasonable questions, desiring that you should just remain in a state of awe and accept whatever they say about the product.

Don't let them fool you: Insist on a clear explanation of the product—technology or otherwise. If you're being told it's "too complex," push back and insist on a simple explanation. If the company or individual persists in their insisting that the technology is "too complex" for you to understand, walk away. It's not magic—it's likely just a computer doing basic computer stuff. Be even more suspicious if there are "experts" perpetuating this same mysterious narrative.

C—2—6

Expert Opinions and Testimonials—Masking the truth is one thing, but companies also use manufactured credibility to sell their mysterious narrative as "expert opinions" and "testimonials." These are carefully crafted third-party proofs that are leveraged to sway public opinion or stamp out skepticism. In many cases, supposed "expert opinions" are the result of incentivized efforts, creating the illusion of credibility—think of all those countless review videos being churned out on YouTube. Sometimes, the "expert" making the review is far from being an expert at all. It's like if your friend hypes up a terrible movie because they got free tickets—don't just blindly trust their review! These endorsements are often used to give the impression that the technology is more widely accepted or proven than it really is. By carefully selecting who speaks on behalf of a product and selectively showcasing manufactured feedback, companies can distort the truth and manipulate potential customers into believing that their products are better than they actually are.

Here's how to avoid the trap: Look up the "experts" behind the review—are they being paid by the company, or do they profit from the product's success? Look for independent reviews instead, but consider many different independent reviews of different rating levels. In doing so, you can look for common points of concern or dissatisfaction. Also, when you can, sleep on it. Don't make impulsive, hasty decisions, but give yourself time to reasonably think over your findings before making a choice. Consider if the proponents are trying to sell you a legitimate, proven technology, or a promise of what could be.

C—2—7

The "We're Already Here" Fallacy—Fake experts create false trust, but this next trick is even more insidious—acting like the future is already here. An incredibly dangerous trick, the "we're already here" fallacy happens when tech companies present the current iteration of their product as if they've already achieved something monumental—when, in fact, they've only scratched the surface of what they're marketing the product as being able to do in the first place. They'll say things like, "We've built a fully autonomous car," when what they've really done is create a system with limited, controlled capabilities. Or they'll claim to have created "true AI," when the reality is that they've only implemented basic algorithms or narrow models. Take OpenAI—they claim ChatGPT "thinks" like a human, but it just remixes data. In essence, the companies or individuals making such bold claims are acting like they've built a car that can drive you to the moon, yet it can't really even make a left turn without stopping mid-turn! This tactic preys on the consumer's desire to believe that the future has arrived and that the technology is already capable of what is promised. What they're selling is usually just an early iteration that cannot live up to all of the marketing hype. I've observed this happening a lot with independent game developers peddling half-baked games on the Steam gaming platform.

Don't fall for it: Compare the claims to reality of the technology's capabilities. For example, if the claim is "fully autonomous," check if it still requires human input or oversight. If it does, it's not actually autonomous at all. If you've been duped by such claims, seek financial recompense—hopefully as a full refund. If necessary, seek the assistance of your state's Attorney General's office.

C—2—8

Now that you know some of the tech industry's tricks, let's wrap up with how to use this knowledge. The tech industry thrives on deception, and the "AI" boom is no exception: vague terms like "AI" and "the cloud," meaningless buzzwords, overblown promises, secretive black boxes, fake experts, and the claim that the future is already here. With these tools, you can spot the lies from a mile away. Start asking questions like: What does this tech really do? What problems does it solve? Are there any existing technologies that were already solving said problems? What proof backs up these claims? Who's providing the proof, and what's in it for them? Don't stop there, though. Take it a step further—share what you learn with others. For example, call out deceptive "AI" hype on social media or on other platforms you trust and help others learn the truth too. The more average consumers who see through these lies, the closer we get to the truth prevailing. Every time you hear a new claim from the tech world, ask yourself: Is this real, or is it just another marketing trick?

Section 3 — A Call to Rationality and Critical Thinking: Encouraging A More Informed Public Discourse

C—3—1

I've already exposed the fraudulent "AI" narrative, yet that's just one con in a long history of tech industry schemes. Once this bubble bursts, they'll move on to the next fabricated revolution—just as they did with "self-driving cars," blockchain technology, and countless other overhyped technologies that failed to deliver. The real problem isn't just one scam; it's the entire way technology is discussed. Public discourse around technology has long been hijacked. Instead of reasoned discussions, it's a flood of marketing hype, misinformation, and corporate jargon meant to keep people in perpetual awe—despite how little they actually understand. We saw it with "self-driving cars" when companies insisted they were "just a year away" from full autonomy—until reality caught up and they quietly admitted they had no idea how to make it work. We saw it with blockchain technology, which was supposed to revolutionize computing altogether but ended up largely being a vehicle for scams and speculation—not that it hasn't proven to have niche use cases. And now, we see it with "AI," where tech companies are selling the fantasy that algorithms can think and create, when all they're really doing is remixing data in ways that sometimes looks impressive but ultimately isn't intelligence. It's time for a better conversation—one that doesn't revolve around hype but instead cuts through the nonsense and focuses on reality.

C—3—2

The technology industry doesn't just build products and services; it builds narratives, and it has the money and influence to ensure that its version of reality is the loudest. Billion-dollar corporations don't need to prove their claims—they just need to repeat them enough times that people stop questioning them. They pay influencers, fund research that "proves" their technology is world-changing, and flood media channels with press releases disguised as journalism. That's how we end up with nonsense like "AI is coming for all our jobs" or "self-driving cars will eliminate traffic deaths," even when there's no real evidence to support these claims. The media plays a massive role in this deception. Instead of investigating tech companies' claims, journalists and bloggers mostly act as their megaphones. Sensationalism gets clicks, and skeptical reporting doesn't, so the press rarely challenges the industry's marketing-driven narratives. Instead of asking hard questions, they publish glowing reviews of half-baked technology, only admitting later—once it's obvious—that they got it wrong. Look at how long it took mainstream outlets to acknowledge that "self-driving" cars weren't anywhere close to working safely, despite clear evidence from the start that the technology wasn't ready. And then there's the public, which, for the most part, takes tech companies at their word. People assume that because a company, like OpenAI or Tesla, has smart engineers and lots of funding, they must be telling the truth about their products. However, that assumption is exactly what these companies rely on. Most people don't dig deeper into the claims they hear—they just accept that any "revolutionary" new technology must be as groundbreaking as its marketing suggests. The result? A completely broken discourse where tech companies control the narrative, the media amplifies it, and the public just absorbs it, largely without question. That's how bad ideas spread and real issues—like illegal surveillance and monopolistic control over digital infrastructure—get ignored.

C—3—3

Now, imagine a world where discussions about technology are rational instead of breathless and promotional. That would mean recognizing that "AI" isn't thinking; it's just a glorified autocomplete system. It would mean questioning every grand promise the technology industry makes instead of assuming that, because something is said to be new, it must be revolutionary. Instead of the media acting as PR firms for tech companies, they'd actually investigate and challenge the claims being made. For example, if OpenAI were to state that their latest model is at human-level intelligence, the response wouldn't be excitement—it would be, "Prove it." We've already seen glimpses of what such rationality could look like, when users on Musk's social media platform X started digging into Tesla's "self-driving" failures. In doing so, they exposed the gap between the hype and reality. When skeptical analysts pushed back on crypto, they revealed how little real-world utility it actually had. This kind of questioning needs to become the norm, not the exception. If we had a discourse driven by rationality and critical thinking, tech companies wouldn't be able to get away with misleading claims as easily. Instead of pushing whatever sells, they'd have to justify their statements with evidence. That alone would be a massive step toward a more honest and informed public conversation about technology.

C—3—4

And while I paint you such an idealistic world, I also acknowledge that fixing the current mess won't happen overnight. However, there are clear steps that can shift the conversation in the right direction. For starters, the technology industry needs to change the way it talks about itself, yet this won't happen without public action. Companies should be forced—through public pressure, media scrutiny, and consistent enforcement of regulations—to explain their products honestly. Instead of sweeping statements like "this will change everything," they should have to clarify exactly what their technology can and can't do. Yet, they won't do this unless they're consistently forced to. Next, the media needs to stop blindly amplifying tech industry narratives. Tech journalism should be about investigation, not free advertising. The same way financial journalists scrutinize Wall Street, tech reporters should be holding companies accountable for their claims. Every time a company makes a bold statement about what their technology will achieve, the immediate response should be: "Where's the proof?" Finally, the public needs to stop being so gullible, especially when it comes to technology. The average person might not have a deep technical background, but that doesn't mean they have to be passive consumers of hype. It doesn't take an engineering degree to ask, "How does this actually work?" or "What's the catch?" In fact, every time someone questions a tech claim—whether in an article, on social media, or even in a conversation with friends—they potentially contribute to shifting the discourse in a positive direction. This is the real corrective force. As long as enough people push back against hype and demand clarity instead of fantasy, the industry will have no choice but to adjust. However, these changes won't happen overnight, and I want to make sure that point is crystal clear. Yet, they are necessary if we want a public discussion about technology that primarily serves the people buying it rather than the companies that are selling it.

C—3—5

Genuine technological advancements may be slowing down, yet the lies being manufactured to prop up the illusion of consistent progress are more aggressive than ever. The way the conversation has been controlled up to this point does not have to remain the status quo—it can be changed. If the public mindset shifts, the technology industry loses its ability to manufacture hype without consistent negative consequences. Narratives can collapse overnight when enough people push back—just look at how quickly "self-driving" cars went from being an "inevitable future" to a perpetual disappointment, or how the crypto industry largely lost traction once skepticism became widespread. The same will happen to the "AI" myth. The more people refuse to accept its deceptive and empty promises, the less power the industry has to define reality on its own terms. By simply reading this book, you've been armed to see through the "AI" scam, but you can things further. The technology industry thrives on unchecked narratives. Every time you call out a lie, refuse to be swept up in hype, or demand evidence before believing grand claims, you make it harder for them to operate in the shadows. Tech companies want a world where they control the narrative without question. It is up to John and Jane Doe—the average consumers—to make sure they don't get it.

Section 4 — Leveraging "AI" Technologies in Writing This Book: An Irony Revealed

C—4—1

We're nearing the end of our time together, but before we dive into my Christian perspective on "Artificial Intelligence" in the final section, I want to take a moment to address something that might seem rather ironic. In writing this book—where I've made a clear case for why "AI" is, at its core, a fraudulent marketing narrative—I leveraged two of the most commonly fraudulently marketed "AI" technologies to do so: ChatGPT and Grok. Yes, you read that correctly. In the process of writing a book that calls out the fraudulent narratives surrounding "AI," I used tools that are often touted as "AI" to assist in writing the very arguments that prove they aren't actually intelligent at all. Now, let's get this clear: While I will repeat over and over again that "AI" is a con job, I'm not saying that the technologies behind the fraudulent marketing are entirely useless. In fact, when used correctly, some of these technologies can serve valuable purposes. Here's the catch, though: Such technologies must be utilized with the understanding that they're not actually intelligent—and in most cases aren't even groundbreaking or revolutionary. They don't understand, reason, or make decisions like a human. What technologies like ChatGPT and Grok do have, however, are complex algorithms that can process vast amounts of data and generate results based on patterns in a very short time. That alone can be useful in many contexts, but it's far from the intelligent behavior that the term "AI" suggests. Moreover, it's prone to errors, lacks true comprehension, and doesn't function independently. My use of these technologies and the manner in which I've used them further proves they're not what the fraudulent marketing and hype claim they are—think about it.

C—4—2

So, then, why did I use these technologies to help write this book? Simply put, after countless hours of testing both, I knew their individual strengths and limitations, and therefore the contexts in which they could each be useful. I didn't expect them to create groundbreaking insights or solve problems on their own. Instead, I used them for tasks like offering a starting point for paragraphs when faced with writer's block, brainstorming, grammar and flow verification (this definitely requires human oversight), and even playing devil's advocate (Grok does a fairly decent job of this)—always with the understanding that the technology I was using was not intelligent. I used my own genuine intelligence but leveraged ChatGPT and Grok to streamline aspects of authoring this book—though, due to their error-prone nature, using them productively often required more effort than doing it myself from scratch.. The point of this book isn't to dismiss the potential usefulness of tools like ChatGPT and Grok, even though they aren't truly "AI." Rather, it's to highlight the importance of presenting them properly, including their capabilities and limitations. They can indeed be helpful, but they need to be marketed and understood for what they are: complex, sometimes powerful tools that still require human oversight and input. They are error-prone (I cannot stress this enough), and their "intelligence" is simply a well-orchestrated illusion. Users need to recognize such limitations and not be misled by the fraudulent marketing surrounding them, or any other technology being mislabeled as "AI".

C—4—3

By revealing this, I hope to reinforce the need for rationality and discernment when dealing with these technologies—without it, they can generate confident nonsense that misleads rather than informs. Yes, they can be useful, but only if understood and leveraged rightly—and that starts with acknowledging that they are not truly "intelligent" and cannot be fully relied upon, especially for critical decisions. Such discernment sets the stage for a deeper truth—a spiritually-guided truth—and in the next section I'll share why I maintain that "AI" will never truly exist, from a Christian perspective.

Section 5 — A Christian Perspective on Artificial Intelligence: Why The Technology Will Never Actually Exist and How Its Pursuit Is Idolatrous

C—5—1

Here we are—we've come to the final section of this book, and just as every good non-fiction book needs an introduction, it also needs an ending. Up to this point, though I have been driven by my role as a Christian leader and mentor, I've largely worn my technology expert hat—to competently expose the truth about so-called "AI" technology. So, for this final section of the final chapter, I'm setting aside the tech expert role and putting on my Pastor J hat (aka my Christian leader and mentor hat). It's no surprise that I don't buy in to the fraudulent "AI" narrative at all, and it's because I know the truth. I've shared that truth with you over the course of this book, and in this section I will provide some additional truth—in the form of a Christian perspective on "Artificial Intelligence" technology and why I believe it will never exist. Of course, I will also substantiate that perspective with relevant Scripture.

C—5—2

From the Christian perspective, which acknowledges God as the Supreme Creator of all, the idea of "Artificial Intelligence" is inherently flawed, because such an idea suggests that machines can be built that can mimic or supersede human intelligence. In other words, that's to say that a unique and special trait that God designed mankind with—intelligence—one that he himself also possesses, albeit at a greater magnitude, can be replicated using hardware components and software created by finite and fallible humans. The contradiction should be clear: intelligence, a divine trait granted by God, cannot be reduced to circuits and code. The Bible teaches that humans are made in God's image (Genesis 1:27), which means that we possess traits and qualities similar to that of our Creator. If that is an accepted fact, then there could never be a machine that could replicate those traits —such as intelligence—as they are a divine gift, not a human invention, and such a machine would undoubtedly undermine God's role as Supreme Creator of all. After all, how truly divine is genuine intelligence if it can be replicated with manmade technology? There seems to be an inherent logical flaw in such an idea, and one that cannot ever be reconciled with some basic truths shared among Christians—that God is divine, and that his power and divine nature are evident in the entirety of His creation. In fact, though he was speaking specifically about God's wrath on unrighteousness, Paul acknowledged as much in his letter to the Romans (Romans 1:20). To me, at least, this all seems rather obvious, but I can understand where many Christians, living in a world largely driven by technology, wouldn't immediately make that connection. In the same way, they wouldn't likely make the connection between believing that "AI" is a legitimate technology—or even a possible technology—and idolatry.

C—5—3

In the most fundamental sense of the word, idolatry is the act of worshipping something or someone other than God—typically associated with images, icons, statues, or even people, with the exception of Jesus in human form (Exodus 20:3; Exodus 32). And while this is certainly one manifestation of idolatry, Scripture reveals that there's more to it than just that fundamental understanding of the sin. In Colossians 3:5, for example, Paul identifies covetousness as also being idolatry. This revelation, then, seems to suggest that the sinfulness of idolatry doesn't inherently reside in the object or person being idolized, but rather in the intention of the person or people doing the idolizing. As with other sins addressed in Scripture —such as adultery and murder—the heart of the matter is focused on intent. In Matthew 5:21–28, Jesus confirms as much when he states that anger (presumably unrighteous anger) towards another will result in the same liability of judgement as murder, and then goes on to explain that even a lustful glance constitutes adultery. Then again, it's worth noting that there is a biblically-supported difference between killing and murdering, which resides in the intention behind the action. Idolatry, then, is ultimately about intentionally misdirecting devotion—when something or someone is given the reverence, trust, or priority that belongs to God alone. The story of the golden calf found in Exodus 32 serves as a textbook example of idolatry: the children of Israel directed their trust and devotion toward a manmade object in place of the living God. Conversely, if someone were to sketch a conceptual image of that golden calf for historical illustrative purposes—without any intention of reverence or exaltation—such an action would not, in and of itself, constitute idolatry. The distinction lies in whether something or someone is intentionally being elevated to the level of God or beyond that level. That said, in Exodus 23:13 God does warn the people not even to mention the names of false gods. While, at face value, this might appear to point to the object or name itself as being inherently sinful, when the warning is taken within its full context it seems to point more-so to people's fidelity to God. It is a warning about the danger of giving even an inch to something that or someone who might draw their devotion away from Him. So, then, again, the issue is not that the object or name itself possesses some inherent sinfulness. This distinction matters in contemporary discussions, particularly when it comes to emerging technologies. Now, I have no doubt that there are some who have already determined where I am going with this, and some may refute my point insisting that tools like "Artificial Intelligence" merely reflect humanity's creative capacity, and by extension, God's own greatness. However, such a view seems to conflate the legitimate use of human ingenuity with a misplaced confidence in man's ability to replicate divine attributes that only God alone has the ability to replicate. Granted, God does give humans creative ability as image-bearers, but not without limits. The story of the Tower of Babel (Genesis 11:1–9) is a clear lesson in what happens when those limits are overstepped—attempting to elevate human ambition to a divine status. There is a categorical difference between building tools that serve practical needs and attempting to fabricate divine attributes that are under the sole authority of God—such as intelligence, life, or moral autonomy. No computer system or other technology, regardless of its complexity, can bear the Imago Dei (the Image of God) or possess the divine attributes that flow from it. When humanity assigns God-like qualities to machines—declaring them wise, autonomous, or capable of moral reasoning—there is a clear cause for alarm. Why? Because this is where idolatry begins: not in the act of creation itself, but in humanity's persistent attempt to usurp the divine authority and uniqueness of the Creator. In fact, I would go so far as to say that this scenario depicts the same hubris that led to mankind's fall in Eden (Genesis 3), and the same folly that led to Babel's collapse. Just because it is now aimed at "Artificial Intelligence," doesn't mean that it will render a better outcome. Based on historical evidence, I would say that the outcome will, indeed, be all too similar to every other instance of mankind attempting to elevate itself to the same level as God—when mankind attempts to "play God". Granted, we cannot fall again since, while in our old-Adam flesh, we're already in a fallen state. Even so, the outcome will still be nothing short of dismal.

C—5—4

When one considers how such a Christian worldview applies to so-called "Artificial Intelligence," they will likely conclude that the fraudulent "AI" narrative is not just a technological falsehood—it's also a blatant spiritual deception, one that attempts to undermine God's exclusive role as the Creator of not just intelligence, but everything that is good, and useful, and right. The ability to create true intelligence requires a divine nature that is equal to that of God's own divine nature, which no human possesses—nor could they ever create such a thing. Furthermore, any attempt to create a machine that genuinely possesses or mimics what is a divinely obtained attribute— intelligence in this case—is entirely rooted in hubris and is, therefore, nothing short of idolatry—

elevating human creations as being equal to or exceeding those of the Creator God. These truths, rooted in Scripture and Christian faith, have served as the foundation of this entire section, and I pray that they have opened the eyes of many of my brothers and sisters in Christ who might have otherwise been deceived by the tempting cloak the world has wrapped this fraudulent technology in. With that, we have come to the end of our time together, and as we now conclude this book, I take comfort in knowing that you have been armed with the truth—the knowledge that proves the term "Artificial Intelligence" has been manipulated and exploited, sold as a revolutionary leap in technology when, at its core, it is nothing more than an international con job (reiterate thesis—check). Throughout this book, I've exposed the fraud behind "AI," using reason, evidence, and even a biblical worldview to spotlight its fraudulent, deceptive nature. I urge you to stand firm in this truth, share it whenever you can, and help foster a world where technological progress is not shaped by deception, but by truth and transparency.

About the Author

Jared A. Snyder, known also as Pastor J, is a technology veteran with decades of experience across a wide array of disciplines in the technology industry, as well as a Christian leader devoted to truth and mentorship. Having dedicated no less than a decade to the study of theology, he has benefited from the knowledge and experience of esteemed minds, such as that of the late Rev. Dr. Rod Rosenbladt—a distinguished theologian and dedicated Christian apologist known for his emphasis on justification by faith. Drawing from a broad life experience, which includes abject poverty and homelessness, Jared strives to bring depth to all his writing, that it might guide his readers toward truth and discernment.

About the Publisher

Synesi Publishing is an imprint of Zoe's Publishing, LC (Est. June 2017), which was created by Jared A. Snyder, also known as Pastor J, in July 2017. While Zoe's Publishing was established to create instructive and inspiring Christian faith-based literature for children, Synesi Publishing was born from a desire to reach beyond those early years—into the hearts and minds of young adult and adult readers alike.

With a focus on insightful, truth-driven, spiritually-grounded nonfiction works, Synesi Publishing carries forward the core focus of its parent company: to publish books that inspire, teach, and build a foundation of faith and truth that lasts a lifetime.